W9-AYX-166

The Craft of Scientific Writing

The Craft of Scientific Writing

Michael Alley

Prentice-Hall, Inc.
Englewood Cliffs, NJ

Prentice-Hall International, Inc., *London*
Prentice-Hall of Australia, Pty. Ltd., *Sydney*
Prentice-Hall Canada, Inc., *Toronto*
Prentice-Hall of India Private Ltd., *New Delhi*
Prentice-Hall of Japan, Inc., *Tokyo*
Prentice-Hall of Southeast Asia Pte. Ltd., *Singapore*
Editora Prentice-Hall do Brasil Ltda., *Rio de Janeiro*
Prentice-Hall Hispanoamericana, S.A., *Mexico*

Library of Congress Cataloging-in-Publication Data

Alley, Michael.
The craft of scientific writing.

Bibliography: p.
Includes index.
1. Technical writing. I. Title.
T11.A37 1987 808'.0665021 86-22534

ISBN 0-13-188855-2

Printed in the United States of America

For two special scientists—
my mother and father

Contents

How This Book Will Help You

In October 1984, the weak writing in a scientific report made national news. The report, which outlined safety procedures during a nuclear attack, advised industrial workers "to don heavy clothes and immerse themselves in a large body of water." The logic behind this advice was sound: water is a good absorber of heat, neutrons, and gamma rays. Unfortunately, the way the advice was worded was weak. Was everyone supposed to be completely submerged? Was it safe to come up for air? Besides being unclear, the writing conveyed the wrong impression to the public. The report came across as saying "*go jump in a lake*"— not the impression you want to give someone spending a quarter of a million dollars to fund your research.

Although chances are slim that Dan Rather will ever read your scientific papers on national television, you as a scientist—physicist, chemist, biologist, or engineer—must assume responsibility for what you write.

Scientific writing is integral to your profession. You must write proposals to secure research funding. You must write progress reports to maintain that funding. And you must write final reports to present the results of your research. Your writing obligations do not diminish as you advance in your career; in fact, they increase. Most top management scientists spend an average of one-fourth of their working day at their desks writing. Moreover, an overwhelming percentage of these scientists claim that their ability to write helped them advance.*

Although scientists spend a great deal of time writing, only a very small percentage of scientific papers and reports

---

*Richard M. Davis, *Technical Writing: Its Importance in the Engineering Profession and Its Place in Engineering Curriculum, AFIT TR 75-5* (Wright-Patterson AFB, Ohio: Wright-Patterson Air Force Base, 1975).

are well-written. Most scientific writing is weak—it does not efficiently communicate the results of the research. Most papers and reports are not well-structured. They contain illustrations that do not mesh with what is written. And much of what is written is imprecise and unclear.

Why is scientific writing so weak? There are many reasons. For one, the subject matter of scientific writing is usually complex. Scientific research is full of complicated vocabularies, images, and ideas that are difficult to express in writing—the more difficult the subject matter, the more difficult the job in presenting it. Another reason scientific writing is weak is that scientists often don't have the time to make their writing strong. Research projects are difficult to schedule and the time allotted for writing often gets crunched. The principal reason, though, that most scientific writing is weak has nothing to do with the complexity of research or the scheduling of projects. The principal reason that most scientific writing is weak is that most scientists *don't* know what strong scientific writing is. Most scientists never receive any formal training in scientific writing. Although they may take an undergraduate class or two in composition or business writing, few scientists ever study the specific problems of scientific writing—problems that arise from presenting the complex vocabularies, images, and ideas of scientific research.

Because most scientists have never sat down and studied scientific writing, many conceptions they have about scientific writing are really misconceptions. This book dispels the common misconceptions about scientific writing and uses examples from actual research papers to show you how to make your writing strong. In this book you will learn how to write clear and precise sentences, how to use illustrations that mesh with what you've written, and how to organize your ideas logically. More important perhaps, you will learn how professional writers actually sit down and write: how they get in the mood, how they write first drafts, how they revise.

The first section of this book explains why most scientific writing is weak. This first section details the problems you face when you write a scientific paper and then introduces you to three stylistic tools—language, illustration, and structure—that can help you solve those problems.

The next three sections of the book show you how to use each of these tools. Unlike many other writing books, this book does not bombard you with simplistic rules or cookbook recipes. Although some questions of style such as grammar and punctuation can be handled with set rules, most stylistic questions cannot. Instead of giving you lists of one-liners, this book uses many examples from actual research papers to show you the differences between strong and weak scientific writing. What this approach does, in effect, is show you how to critique scientific writing. The first step toward becoming a strong writer is to become a strong reader. These three sections help you take that step.

The final section of this book covers home plate; it details the actual process of sitting down to write. No matter how much technique you know, you're pretty much worthless as a writer if you can't sit down and write. In this section, you will learn how professional writers schedule their writing time. You will learn ways to get into the mood—what exercises to do, what foods to eat. This section will run you through the gauntlet of writing the first draft and show you effective techniques for revision. This section begins with a blank piece of paper and finishes with a completed manuscript.

I wish I could tell you that this book will make your writing easy. But that's not the way writing is. Writing is hard work. The best writers struggle with every paragraph, every sentence, every phrase. They must write, then rewrite, then rewrite again. Scientific writing is a craft, a craft you must constantly hone.

PART ONE

Scientists and Scientific Writing

CHAPTER 1

Misconceptions: The Principal Source of Weak Scientific Writing

We are all apprentices in a craft where no one ever becomes a master.

Ernest Hemingway

Scientists—biologists, chemists, physicists, and engineers—have a split personality about scientific writing.

As readers, scientists are harsh critics. They will state that most scientific writing is poor. Many scientists point to organization as the main problem. Either the main results have been buried in the writing or else the writing skips from one idea to another in a Brownian motion. Other scientists raise more specific criticisms, such as unclear illustrations or pretentious word choices. As readers, scientists assume an idealistic pose about the role of scientific writing. They see the need for more efficient communication in science. Too many poorly written papers are being published, they will say. Our writing must be clear and concise.

As writers, however, scientists show a second personality. No longer are they the outspoken critics of style. Instead, they turn defensive, cringing at constructive criticism and balking at requests for needed revisions. As writers, scientists shun their idealistic visions about the role of scientific writing. They close their eyes to the need for more efficient communication in science and concentrate instead on their own research deadlines, the pressures on them to publish, the number of reports they have to write. If reminded of their earlier idealistic statements about writing, many scientists will fumble for excuses. Some will even recant.

Why do scientists have this split personality?

The first personality—the harsh critic—is fairly easy to understand. No matter how strong the writing is, most scientific papers are inherently difficult to read. The subject matter is filled with complex images. Many of these complex images are abstract: electromagnetic field configurations, quantum orbits of electrons. Other images are detailed, such as the nuclear fusion experiment shown in Figure 1-1.

Another aspect that makes reading scientific papers difficult is the general appearance of the language. Scientific language is full of specific terms, abbreviations, and odd hyphenations:

> *This paper shows how intensity fluctuations in the frequency-doubled output of a Nd:YAG pump laser affects coherent anti-Stokes Raman spectroscopy (CARS) signal generation.*[2]

"Frequency-doubled," "Nd:YAG," "anti-Stokes," "CARS"—these aren't the kinds of expressions you run across in the morning paper. They are part of a different language—the language of lasers. In science, there are many such languages; languages that are constantly evolving.

Figure 1-1. Nuclear fusion experiment. In this experiment, an accelerator focuses lithium ions onto deuterium-tritium pellets in an attempt to produce nuclear fusion.[1]

Besides complex words and abbreviations, scientific language also contains many mathematical equations. Equations slow your reading; you must stop and work through the physical meaning of each variable.

The burning rate (Y) of a homogeneous solid propellant is given by

$$Y = \frac{\rho}{\alpha}(2\lambda\hat{\tau})^{1/2}\,\mathrm{Le}^{n/2}\left[\frac{c(1-\sigma)}{(c-\sigma)(1+\gamma_s)}\right],$$

where ρ*,* c*, and* λ *are the gas-to-solid ratios for density, heat capacity, and thermal diffusion,* $\hat{\tau}$ *is a scaled ratio of the rate coefficient for the gas-phase reaction,* γ_s *is the ratio of the heat of sublimation to the overall heat of reaction,* Le *is the gas phase Lewis number, and* α *is the fraction of propellant which sublimes and burns in the gas phase.*[3]

Although complex images and complex language make reading scientific papers difficult, the most taxing aspect of reading scientific papers is compression. Scientific writing is compressed. Most journals impose tight length restrictions on papers. Therefore, scientists must squeeze descriptions of complicated theories and experiments into a few paragraphs, sometimes a few sentences. This compression thickens the writing and greatly increases the concentration required of readers.

> *Engine knock is the mechanical response (vibration) of an engine to a pressure rise in the combustion chamber, a rise so rapid that the pressure is no longer uniform through the chamber. In our experiment, the end gases in a high-swirl, homogeneous-charge research engine were isolated in the center of the cylindrical combustion chamber by simultaneous ignition at four spark plugs equally spaced in the circumference of the cylinder wall. Manifold pressure and temperature were used to control the fraction of fuel undergoing autoignition. The residual gas fraction in the low-compression-ratio engine was reduced to less than one mole percent of the total charge by igniting every third cycle. The result is an unusually low cycle-to-cycle variation even when knock is occurring.*[4]

Because scientific papers and reports are so complex, writing mistakes—even small ones—can confuse readers.

> *Antibodies are formed soon after which the lymphocytes quit reproducing.*

Are antibodies formed after the lymphocytes quit reproducing, or will the lymphocytes stop reproducing after antibodies are formed? Standing alone, this sentence puzzles readers. However, in the middle of a scientific report, this mistake annoys readers. Readers must stop and reread the previous paragraph, figure out what the writer meant, regain concentration, and move on. This writing mistake is small; it probably costs readers only a couple of minutes. Unfortunately, these mistakes in scientific writing often accumulate:

> *The object of the work was to confirm the nature of electrical breakdown of nitrogen in uniform fields at relatively high pressures and inter-electrode gaps which approach those obtained*

> *in engineering practice, prior to the determination of the processes which the criterion for breakdown in the above-mentioned gases and mixtures in uniform and nonuniform fields of engineering significance.*

This sentence bewilders. Not only does it present too many details, but it presents them in an unclear way. What exactly does it mean to "confirm the nature" of something? And what exactly are "fields of engineering significance"? Worse yet, the transition phrase "prior to the determination of" makes no transition; it leaves a hole in the middle of the sentence. This kind of writing angers readers. As if the subject matter isn't difficult enough, readers must wade through weak writing. When this kind of paragraph finds its way into a poorly organized paper, readers can spend hours piecing together information. Those hours turn scientific readers into harsh critics of writing.

The second personality—the defensive writer—is somewhat perplexing. Scientists are not naturally defensive about their work. Most scientists, especially the good ones, readily accept challenges to their theories and experiments. They view such challenges as healthy. Why then are they so defensive about their writing?

Is it pride? For a few scientists, yes, but not for most. In their research, most scientists depend on the principle of trial and error. Because scientists aren't too proud to admit mistakes in their research, they shouldn't be too proud to admit mistakes in their writing.

Is it fear then—the fear of having to rework and rewrite? Writing is hard work, lonely work; yet by the time most scientists are ready to write, they have already spent months, sometimes years, doing research. Surely, they would not hesitate to spend a few extra days redrafting a paper if they believed the work was necessary.

So why are most scientists defensive about their writing?

The reason that most scientists are defensive about their writing is neither pride nor fear; it is confusion. Most scientists are honestly confused about how to write scientific papers. They have never sat down and thought out exactly why they write or what they want their writing to do. Instead, they rely on a set of vague conceptions they have about scientific writing; conceptions they've developed over many years; conceptions about what scientific writing is supposed to look like; how it should be approached. These vague conceptions don't stem from any formal training in scientific writing or any calculated thought on the subject. Rather, these conceptions arise from three untrustworthy sources: scientific folklore, rules that scientists remember, and examples that scientists read.

Science, like every other profession, has its own folklore. The folklore of science is not spread around campfires or wood stoves, but in smoke-filled computer rooms and in dark laboratories with black plastic on the windows. There is folklore about mathematics: Mathematics is just a tool. There is folklore about doing experiments: Never leave an experiment that is working. The folklore about scientific writing does not address particular stylistic aspects; instead it reflects the general attitudes that scientists have about writing.

> Scientific writing is a mystical science.
>
> Scientific writing is not really that important.
>
> Scientists can't write; if they could, they wouldn't be scientists.

This folklore about scientific writing is simplistic. More than that, this folklore is untrue; it skews the approach that scientists take to their writing.

The second source for the vague conceptions that scientists have about writing comes from rules that scientists remember. If you ask a scientist to advise you about writing, you're likely to get a quick list of rules.

> Be objective.
>
> Use synonyms for variety.
>
> Begin each paragraph with a thesis statement.

These rules come from freshman composition classes taken years ago; late night conversations with thesis professors; DOs and DON'Ts articles cut out of company newsletters. These rules are simple and general; that's why they're easy to remember. However, when it comes time for you to actually sit down and write a scientific paper, these rules don't help much. What makes these rules easy to remember also makes them valueless; they're too simple and general. Let's face it: Writing scientific papers is difficult, even more difficult than reading scientific papers. The writing problems in scientific papers are much too complex to be solved by a list of one-liners. Except for grammar and punctuation, the stylistic elements in scientific writing are not constant. The way you organize paragraphs, the words you choose, the kinds of verbs you use (active or passive)—these decisions depend on your particular situation. They depend on your research and your audience.

Moreover, most rules that scientists remember don't really affect the way they write. Most rules that scientists remember are imaginary crutches. The biggest influence on how scientists write comes from other scientific papers they've read, not from any rules they've remembered. Just as what you hear influences the way you speak, what you read influences the way you write. Word choices, sentence rhythms, even the ways that papers are organized are consciously (and subconsciously) absorbed by readers. Unfortunately, most writing examples in scientific literature are weak. Most scientific papers are not well structured, the language is imprecise and unclear, and the illustrations do not mesh with the words.

The principal reason that most scientific writing is weak is not that scientists can't write. Most scientific writing is weak because scientists don't know what strong scientific writing is. Many conceptions that scientists have about scientific writing are really misconceptions, and these misconceptions confuse scientists, making them defensive about their writing.

Perhaps the most common misconception has to do with the way that scientists approach scientific writing. Many scientists look at scientific writing as a mystical science. Scientific writing is not a mystical science. For one thing, scientific writing is not a science. It contains no proven theories or sophisticated experiments. Scientific writing is a craft. It is a skill that must be developed through practice. Moreover, scientific writing is not mystical. In fact, scientific writing is very straightforward. Unlike other forms of writing, such as fiction where the goals are impossible to define, scientific writing has a single objective: to inform the reader. You, the writer, must present your research in a way that your audience can understand. There's nothing mystical about that.

Another common misconception deals with the importance, or lack of it, that scientists assign to scientific writing. Many scientists feel that scientific writing is not that important because the skills needed for scientific writing aren't as sophisticated as those needed to do the research. Perhaps the intellectual demands for writing a scientific paper aren't as great as those for splicing genes or solving nonlinear differential equations, but *communicating* results is just as important as *obtaining* results. What progress have you made if you design a new X-ray laser but can't communicate the design to anyone with the purse strings to invest in it?

The relationship of scientific writing to scientific research is much like the relationship of putting to golf. Researching a particular problem is analogous to reaching the green. The research involves the most complex strokes: driving the fairway, pitching over water, chipping out of the sand. Once you are on the green, you have obtained results. Communicating those results then is the putting. A putt requires neither the power of a drive nor the sophisticated swing of an iron shot, but it does require a certain skill; a skill that is important.

A third common misconception concerns the way that scientists see themselves as writers. Many scientists believe that there's something innate that prevents them from ever writing successfully. These scientists assume that if they had the skills to write they wouldn't have the skills to be scientists. This misconception is ludicrous. Although there may be some physiological correlations between being adept at analytical skills and weak at verbal skills, scientific writing is not a some-have-it, some-don't craft. Anyone who can perform sophisticated scientific research can write a successful scientific paper. Scientific writing is not an "inspiration" craft; it's a "perspiration" craft. Don't think though that becoming a successful scientific writer is a one-night affair. Great writers never cease learning.

Writing a scientific paper is much like shaping a piece of metal on a lathe. When you work on a lathe, you shape the metal to fit into a particular machine. In scientific writing, you shape your papers to inform a particular audience. Your first few drafts of a paper are rough cuts on the metal stock. Slowly, your metal piece takes shape so that in your final passes you are only skimming off a few thousandths. In your final few drafts of a paper, you are doing the same: rewording a sentence here, strengthening a transition there.

REFERENCES

1. Pace VanDevender, "Ion-Beam Focusing: A Step Toward Fusion," *Sandia Technology*, 9, no. 4 (December 1985), pp. 2–13.
2. L. A. Rahn, R. L. Farrow, and R. P. Lucht, "Effects of Laser Field Statistics on CARS Intensities," *Optics Letters*, 9 (1984), p. 223.
3. S. B. Margolis and R. C. Armstrong, "Two Asymptotic Models for Solid Propellant Combustion," *Combust. Sci. Technol.*, 47 (1986), p. 1.
4. J. R. Smith, R. M. Green, C. K. Westbrook, and W. J. Pitz, "An Experimental and Modeling Study of Engine Knock," *Twentieth Symposium (International) on Combustion*, (Pittsburgh, PA: The Combustion Institute, 1985).

CHAPTER 2

Style: The Key to Strong Scientific Writing

The greatest merit of style, of course, is to have words disappear into thoughts.

Nathaniel Hawthorne

Science is an evolving cordillera of knowledge with many different branches and chains. As scientists perform new experiments and develop new theories, mountains in the cordillera evolve. Peaks grow; chasms and ridges develop; sometimes new mountains arise. Beneath this cordillera there lie substrata of knowledge. Every study in science is built upon these substrata. For example, plasma physics rests on the substrata shown in Figure 2-1.

In this cordillera analogy, scientific papers are maps for understanding the research of scientists. A successful paper will lead its readers to the new level of understanding attained by the research; an unsuccessful paper won't. The *style* of a paper is the map's trail. The style determines whether the paper is successful. There are three principal elements of style in scientific writing: *language, illustration,* and *structure.*

The first two elements—language and illustration—determine the condition of the trail. Is the trail rocky or smooth? Does it trip the reader? Language encompasses not only choice of words but also how words are used. Are they precise? Are they clear? Language includes the way words are arranged in sentences; the way sentences are arranged in paragraphs; and more. Language is the use of numbers, equations, symbols, and abbreviations; it is the use of examples and analogies.

Illustration is the meshing of words with pictures. There are two kinds of illustrations: tables and figures. Tables are arrangements of words and numbers in rows and columns. Figures are everything else: photographs, drawings, diagrams,

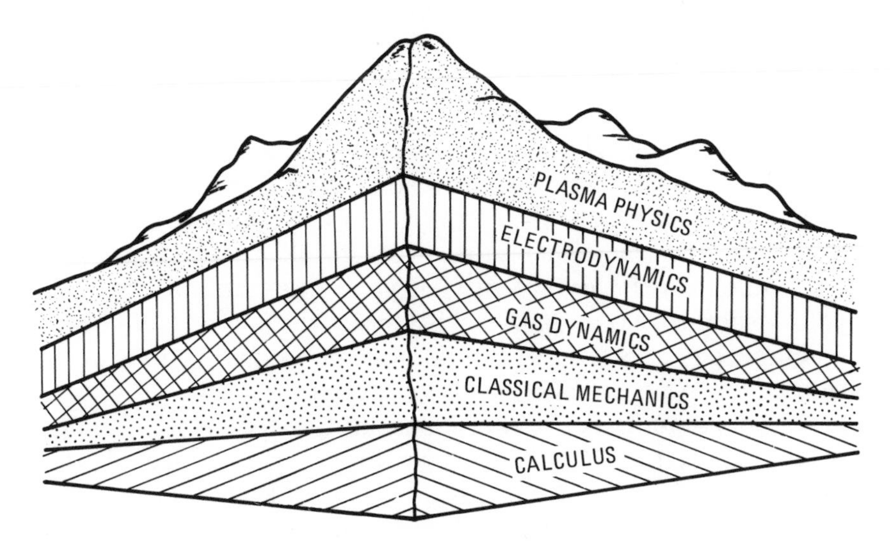

Figure 2-1. Substrata of plasma physics

and graphs. Strong scientific writing demands more than just the presence of illustrations. Your illustrations must mesh with your language. Do your figures have the same precision as your prose? Are your readers left wondering about certain details in your tables?

The third element of style, structure, is the direction of the paper. At what elevation does the trail begin? Does the writing path go straight up the face of a cliff or does it wind along ridges? Structure is the strategy of scientific writing.

There are two common misconceptions associated with style. First, many scientists mistakenly think that style is the same thing as format. Style is the way your research is arranged within a paper or report: the way you word sentences, the complexity you give to illustrations, the results you choose to accent. Format, on the other hand, is the way your paper or report is arranged: the typeface used, the way pages are numbered, the way sources are referenced.

There is no absolute ordained format for scientific writing. Because most scientific journals, laboratories, and corporations establish their own formats, these formats vary. (For some general guidelines regarding format, see the format references listed in the bibliography.) Many scientists fret over the different formats in scientific writing. They can't understand why *Journal A* uses one type of referencing system while *Journal B* uses another, or why one laboratory capitalizes "X-ray" and another laboratory doesn't. These scientists seek absolutes in scientific writing; they mistakenly treat scientific writing as a science instead of a craft. Instead of worrying about format, over which you have little control, you should worry about style, something you do control. You should worry about your word choices, the complexity of your illustrations, the way you structure your papers. Stylistic decisions determine whether readers will understand your research. Stylistic decisions determine the success of your writing.

Many scientists also hold a second misconception about style, namely that in scientific writing everyone's style is the same. Style is individual. Just as no two hikers would chart the exact same path up a rugged mountain, no two scientists would present a new theory or experiment the exact same way. You not only shape your style with small steps such as word choices, sentence rhythms, and analogies, but also with large steps such as organization, the background material you provide, the way you accent results. Given the same research and audience, there are usually several different structures—significantly different—for writing a successful paper. The existence of more than one strategy for a paper bothers many scientists. Many scientists search in vain for absolute solutions to their writing problems. Again, these scientists mistakenly treat scientific writing as a science instead of a craft.

Just because style is individual does not mean that you have a license to write as you please. In scientific writing, your style must meet two constraints:

1. You must inform your audience as efficiently as possible.
2. You must stay honest.

First, you must inform your audience as efficiently as possible. In scientific writing, "efficiency" does not mean the paper with the shortest length; rather, the paper that takes readers the shortest time to understand.

Second, you must be honest. Being honest means including all data points, even those that don't fit the curve. It means not hiding flaws in your research beneath a snowbank of complex writing. To be honest, you must give fair treatment to opposing theories and experiments. Science is not religion. You must base all your arguments on logic, not emotion.

These two constraints separate scientific writing from other types of writing. Many scientists mistakenly think that all kinds of writing are basically the same. Although there are many similarities between different types of writing, scientific writing has its own constraints and problems. Scientific writing is not the same as literary writing. In scientific writing, you have a single defined purpose, that being to inform a particular audience about a particular research project. Literary writing, on the other hand, has no single defined purpose. There can be any number of reasons behind the writing of a poem or story: anger, love, depression, even boredom.

Scientific writing is not the same as journalism, either. In journalism, you generally write every article for the same audience. You make the same assumptions about what your audience knows and why your audience reads your article. In scientific writing, though, you must reconsider your audience every time you write. In one paper, your readers may be experts on your research mountain. In another paper, your readers may be nonscientists who have never even heard of your research mountain.

Scientific writing is not the same as business writing. In business writing, you write to advance your business. In scientific writing, you write to advance science. Sometimes these purposes concur, but often they don't.

Although you can learn much by studying the writing of other disciplines, you must remember that scientific writing has its own constraints and problems. What may be good advice for the fiction writer may prove disastrous for the scientific writer.

INFORMING YOUR AUDIENCE

The purpose of scientific writing is to inform. Scientific writing serves the audience, not the writer, and audiences for scientific papers are diverse. Besides stirring the interest of scientists on your own mountain, your research may also interest scientists on other mountains as well as nonscientists interested in funding your work.

Let's say you helped develop an implantable electronics device that delivers insulin to the bloodstream. You might write one paper for a symposium of electronics engineers, who are familiar with integrated circuitry, but know little about diabetes or the demands placed on circuitry implanted in the human body. You might then write a second paper for a journal of medicine, whose readers are familiar with diabetes, but who don't know the difference between analog and bipolar switches. You might present the same research yet a third time to a general science journal whose readers have a variety of backgrounds.

The same research, but three different audiences and three different papers.

It is important to present your research to various audiences. You can't be content with communicating your results to cronies on your own particular ridge. You must communicate your research to scientists on other mountains, even mountains in different branches and chains of the cordillera. For example, the electronics technology for the implantable insulin device came from weapons research.[1]

Often, you must communicate your work to people who hold the purse strings for your research. Typically, these readers will be nonscientists interested not so much in how your research works, but what importance your research holds. Therefore, to secure your funding, you must tell nonscientists what your research is and convince them of its importance. Knowing how to communicate your research to nonscientists is not only important for securing funds. As Einstein said,

> *It is of great importance that the general public be given the opportunity to experience—consciously and intelligently—the efforts and results of scientific research. It is not sufficient that each result be taken up, elaborated, and applied by a few specialists in the field. Restricting the body of knowledge to a small group deadens the philosophical spirit of a people and leads to spiritual poverty.*[2]

Informing audiences—both scientific and nonscientific—presents many stylistic problems for scientists. There are problems with language. Throughout the cordillera of science, specific vocabularies have evolved.

> *After many thermal decomposition studies of hexahydro-1,3,5-trinitro-s-triazine (RDX) and octahydro-1,3,5,7-tetranitro-1,3,5,7-tetrazocine (HMX), there are still unresolved questions, such as whether decomposition occurs in the solid or gas phase, whether the initial step in the decomposition involves the breaking of a C-N or an N-NO_2 bond, and whether autocatalysis plays a role in the decomposition process.*[3]

Not only are these vocabularies full of unusual words such as "5-trinitro-s-triazine," but many words take on special meanings when used on particular mountains of research. For example, a paper titled "Effects of Humidity on Avalanche Growth and Streamer Initiation" may call to mind all sorts of bizarre images to readers not familiar with the vocabulary of gas discharge physics.

Informing audiences also presents illustration problems for scientists. An illustration such as the solar energy experiment shown in Figure 2-2 raises questions for readers. What are the streaks of light on either side of the tower? Why are some of the mirrors turned upside down? Scientists must answer these questions in their writing; otherwise, readers will come away unsatisfied.

Finally, problems with structure arise as scientists try to inform audiences about complicated research. Scientific research is built on many layers of knowledge, and although your audiences will be familiar with most substrata—mathematics, basic chemistry, basic physics—they will often be unfamiliar with your mountain's top few layers of theories and experiments. Therefore, you must structure your paper to guide your readers through these top layers. This burden does not rest on your audience; it rests on you, the writer. Before you begin any scientific paper or report, you must ask yourself two questions:

1. Who is going to read this paper?
2. Why are they going to read it?

The answers to these two questions will affect the way you shape your language, what kinds of illustrations you choose, and how you structure your papers. In scientific writing, audience influences style, perhaps more so than in any other type of writing.

Because scientific research is so complex and the audiences are so varied, serving your readers will not be easy. Most stylistic decisions will not be clear-cut. Many times, you must choose a difficult path. Just remember that there is no perfection in scientific writing. Scientific writing is a craft, not a science.

Figure 2-2. Solar One Power Plant located near Barstow, California. Mirrors focus solar energy onto a central receiver where water is converted to steam. This steam powers a turbine to produce 10 megawatts of electrical power to a utility grid.[4]

STAYING HONEST

In scientific writing, staying honest means presenting all your data points, even those that don't fit the curve. It means not burying cracks in your research beneath a snowbank of complex writing. If you know your vacuum pumps have coated your experiment with mercury vapor and you suspect that vapor has influenced the results, then you would be dishonest if you did not state your suspicion. As Linus Pauling said, "Science is the search for truth." If you hide the truth, then you have not advanced science. Writing that hides the truth is not scientific writing.

Staying honest also means distinguishing conjecture from fact. Suppose that you studied why the Solar One Power Plant (in Figure 2-2) did not meet its power production goal. Suppose that you suspected the volcanic eruption of Mexico's El Chichón as the main reason—the volcano possibly spewed enough smoke into the upper atmosphere to reduce sunlight to the plant. If in your report you included a sentence such as

The volcanic eruption of El Chichón caused the Solar One Power Plant to fall short of its power production goal.

you would be dishonest because you would be stating a conjecture as fact. To be honest, you must qualify that statement.

A possible reason for the Solar One Power Plant missing its power production goal was the volcanic eruption of El Chichón.

Staying honest also means giving fair treatment to opposing theories and experiments. In much of science, there is disagreement as to how results should be interpreted. Disagreement is healthy; it pushes scientists to refine their thinking. For example, many scientists believe that the greatest contributor to quantum mechanics was not Bohr or Heisenberg who helped lay its foundation, but Einstein who argued so vigorously against it. In your arguments, however, you must be fair—you cannot take any cheap shots. You must base all your arguments on logic, not on emotion.

Staying honest is not the same as being objective. Many

scientists mistakenly believe that scientific writing must be objective. To be objective in scientific writing, you must honestly report how nature acts, then explain those acts with an impartial eye. Although objective scientific writing is an admirable goal, it is impossible to attain. Why?

For one thing, when you write your research, you don't retrace every step you took when you performed the research. You don't walk your readers through every vacuum leak or every computer error. If you tried to, your papers would be too long for anyone to read. In scientific writing, you reshape your research into a trail for your readers. Scientific papers are not diaries. A diary account of scientific research would lead readers down gorges and gullies. Scientific papers are shaped trails that are always ascending; and you perform the shaping. You decide what to tell readers and what to leave out, and these decisions are *subjective* decisions.

Also, when you present research, you qualify your facts. You decide whether an experimental error is "neglible" or "small" or "tolerable" or "significant." Each qualifier tints your research a different way. The point is that you choose the qualifiers. Again, you make subjective decisions. How do you remain objective in your writing? The truth is you can't remain totally objective, but you can be honest.

REFERENCES

1. Gary Carlson, "Implantable Insulin Delivery System," *Sandia Technology*, 6, no. 2 (June 1982), pp. 12–21.
2. Albert Einstein, Foreword in *The Universe and Dr. Einstein*, by Lincoln Barnett, (New York: Sloane, 1948), pp. 1–2.
3. Rich Behrens, "Nitramine Thermal Decomposition by Simultaneous Thermogravimetry Modulated Molecular Beam Mass Spectrometry," *Sandia Combustion Research Program Annual Report*, (Livermore, CA: Sandia National Laboratories, 1985), chap. 5, p. 2.
4. L. G. Radosevich, *Final Report on the Experimental Test and Evaluation Phase of the 10 MWe Solar Thermal Central Receiver Pilot Plant*, SAND85-8015, (Livermore, CA: Sandia National Laboratories, 1986).

PART TWO

Language

CHAPTER 3

The Way That Words Are Used

It is impossible to dissociate language from science or science from language, because every natural science always involves three things: the sequence of phenomena on which the science is based; the abstract concepts which call these phenomena to mind; and the words in which the concepts are expressed. To call forth a concept, a word is needed; to portray a phenomenon, a concept is needed. All three mirror one and the same reality.

Antoine Lavoisier

In writing, language is the way that words are used. Language is word choice, the arrangement of phrases, the structuring of sentences and paragraphs, and more. In scientific writing, language includes the use of numbers, equations, and abbreviations; it includes the use of examples and analogies.

Your use of language must meet the two constraints of style in scientific writing. To meet these constraints, your language must first be *precise*. You must say what you mean.

Your language must also be *clear*. While precision means choosing the right words, clarity means not choosing any wrong ones.

Besides being precise and clear, you should anchor your language in the *familiar*. It is important to use language familiar to your readers in scientific writing. Before readers can understand anything new, they must see it in relation to something they already know. You should also be *forthright* in your language. When you write a scientific paper, you become a teacher, and as a teacher you want to convey a sincere and straightforward attitude. You shouldn't be pompous; nor should you be arrogant.

Because scientists produce so much writing every year, they have an obligation to make their language *concise*. Every word should count. Although you should make your language tight, you should also make it *fluid*. Fluid writing is smooth writing; writing with transition; writing that moves from sentence to sentence, paragraph to paragraph, without tripping or tiring the reader. Finally, your language should be *imagistic*. Your language should show—not just tell—readers what happened in your research.

These goals for language do not carry the same weight. Precision and clarity are the most important. The relative importance of the other goals depends on your audience and research. For instance, in a paper where your research is abstract, say a quantum mechanics paper, being imagistic might become more important than being concise. In that paper you may want to add analogies to insure that your readers understand your ideas.

Don't assume that these seven goals for language are in constant conflict. Most of the time, these goals reinforce one another, as shown in Figure 3-1. When you are clear and forthright, conciseness follows. When you are precise and familiar, clarity follows. Also, don't think that by pursuing these seven goals you will lose your individuality as a writer. Within these seven goals there is an infinite amount of variation. Pursuing these language goals will not cause your writing to sound like everyone else's; rather, it will make your writing successful.

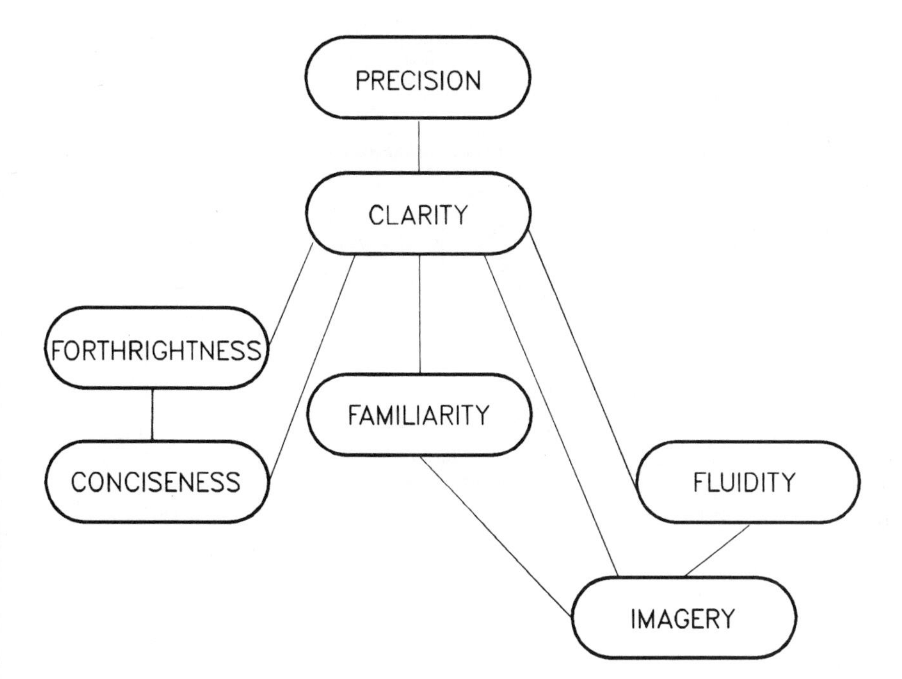

Figure 3-1. The seven goals of language in scientific writing

CHAPTER 4

Being Precise

Use the right word, not its second cousin. The difference between the right word and the almost right word is the difference between "lightning" and "lightning-bug."

Mark Twain

In scientific writing, precision is the most important goal of language. If your writing does not communicate exactly what you did, then you have changed your research. Scientists are no strangers to the importance of precision. Much scientific research involves obtaining high degrees of accuracy in calculations and experiments. In scientific writing, however, precision takes on a different meaning. You must strive not for the highest degree of accuracy, but rather the appropriate accuracy for your research and audience.

CHOOSING THE RIGHT WORDS

As Twain says, word choices are important. You should not choose the word "weight" when you want "mass." You shouldn't choose the word "comprise" when you want "compose."

Air is comprised mainly of nitrogen and oxygen.

"Comprise" literally means "embrace." Therefore, using "comprise" in this sentence is imprecise writing. The word you want is "composed."

Air is composed mainly of nitrogen and oxygen.

Or better still,

Air is 78% nitrogen, 21% oxygen.

Many word choices are difficult; you need one particular word and no other word will do. Despite what you may have learned, few words—if any—are exact synonyms. Some words have similar meanings, yet are not interchangeable. Consider, for example, these four words from the vocabulary of gas discharge physics:

Electrical Breakdown—the transition of a gas from a perfect insulator to a conductor. As a conductor, a gas can assume one of three steady states shown in Figure 4-1; a Townsend discharge, a glow discharge, or an arc discharge.

Spark—the transient irreversible event from one steady state of the electrical breakdown process to another (example: the transition from a glow discharge to an arc).

Gas Discharge—any one of the three steady states of the electrical breakdown process.

Arc—a specific type of gas discharge characterized by low voltage and high current (example: lightning stroke).

To use these four words as synonyms is imprecise language. The first two terms are transitions. The third and fourth terms are steady states. An electrical breakdown is always a spark, but a spark is not always an electrical breakdown. Likewise, all arcs are gas discharges, but not all gas discharges are arcs. Each term has its own specific meaning. Nevertheless, many physicists toss these terms around as synonyms in their writing.

The last decade has seen a rapid development of new techniques for studying the enormously complex phenomena associated with the development of sparks and other gas discharges.

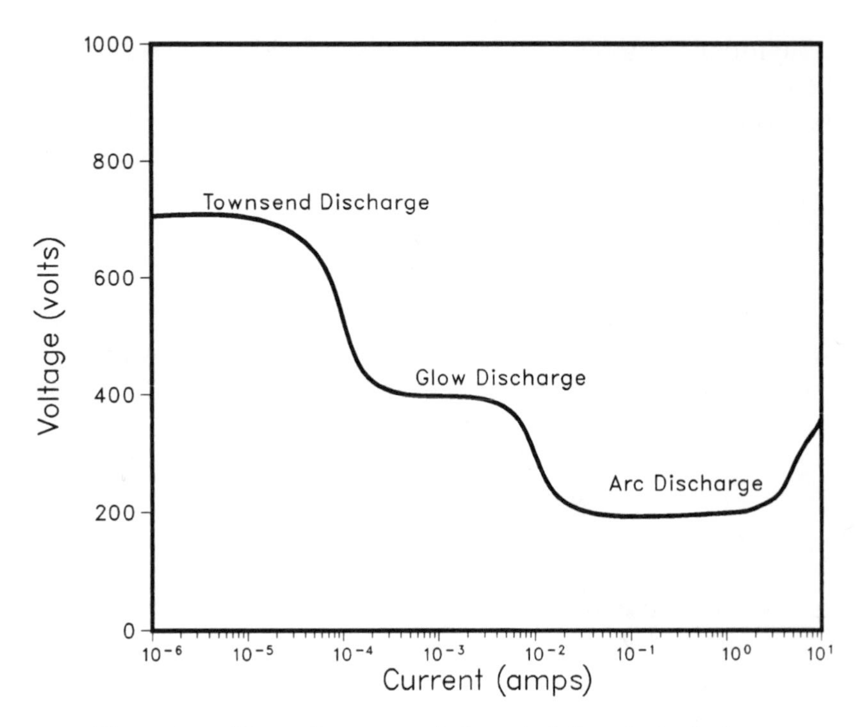

Figure 4-1. The voltage-current characteristics of a typical gas.

Because a "spark" is not a type of "gas discharge," this sentence is imprecise.

Many scientists hold the misconception that using synonyms is a mark of a good writer. These scientists write with a pen in one hand and a thesaurus in the other. Well, synonym writing is not strong writing. Even when you find true synonyms, using them often confuses readers.

> *Mixed convection is a combination of natural and forced convection. Two dimensionless numbers in correlations for mixed convection are the Grashof number and the Reynolds number. The Grashof number (for free convection) is a measure of the ratio of buoyant to viscous forces, and the Reynolds number (for forced convection) is a measure of the ratio of inertial to viscous forces.*

Why did the engineer use "free convection" instead of "natural convection" in the third sentence? This synonym substitution added nothing to the discussion and only served to confuse readers unfamiliar with the vocabulary of thermodynamics.

Most professional writers don't hesitate to repeat a word if that word's the right word. Moreover, most professional writers don't rely on thesauruses. Professional writers rely on dictionaries because dictionaries tell you the differences between words. Dictionaries help you find the right word.

I once had an English teacher who made us write five-sentence paragraphs about different kinds of subjects—subjects such as whales. In the paragraph, whales had to be the subject of every sentence, but you couldn't use the word "whales" more than once. You couldn't use pronouns, either. You had to come up with four synonyms for "whales." Which students did she praise? The students who gave up and decided anything but whales was imprecise? No. She fell all over herself for the freckle-faced boy in the front row who wrote "black ellipsoids of blubber" and "king and queen mammals of the sea." With an evil eye for the rest of us, she said, "Now, he's a writer."

I have news for her: "black ellipsoids of blubber" is no kind of writing; not scientific writing, not journalism, not fiction. You should test any writing advice you've received against the writing of great writers. Did Flannery O'Connor use a string of synonyms in her writing? Did Churchill? Did Einstein? No. They, like Twain, used the right word and only the right word. If a particular word was called for more than once, they repeated that word. You should do the same. In scientific writing, precision is your most important goal.

When choosing words, you must worry about *connotations* (associated meanings) as well as *denotations* (dictionary meanings). Consider, for example, the word "adequate." "Adequate" means enough for what is required. However, would you accept experimental data from a vacuum chamber that was described as "adequate"? No. Why not? Because "adequate" has a bad connotation. When something is described as "adequate," you think of that something as being insufficient. The connotation of "adequate" has come to mean the exact opposite of its denotation.

A common mistake in scientific writing is to use words or phrases that have the correct denotation but wrong connotation. Words and phrases with wrong connotations trip readers because they suggest things that were never intended:

> *The electrical breakdown of a gas is a* dramatic *event in which . . .*

"Dramatic" is not the right word in this sentence. A secondary definition in the dictionary might justify the denotation, but "dramatic" has the wrong connotation; it suggests that electrical breakdown is a theatrical production, a suggestion that distracts readers from the subject matter. Another example:

> *Our laser program* focuses *on the study of . . .*

Again, the denotation of "focuses" is justifiable in this sentence, but the connotation of "focuses" turns the sentence into a bad pun that undercuts the authority of the writer.

BEING SPECIFIC

To inform, you must show. To show, you must provide specific details. Precision in scientific writing means being specific. Don't tell your readers

> *After recognizing the problems with the solar mirrors, we took subsequent corrective measures.*

What were the problems with the mirrors? What were the solutions? This sentence raises questions but does not address them. A stronger sentence would have been

> *After finding that high winds (and not hail) had cracked the solar mirrors, we began stowing them in a horizontal position during thunderstorms.*

Although this second sentence uses a few more words, it tells readers something concrete, a detail they might remember. Take a lesson from fiction writing. Good fiction writers rely on specific details to create scenes because they know that specific details are what readers remember.

Being specific means anchoring general words with details and examples.

With our cathode-ray tube, we measured fast electrons in the arc discharge.

What does "fast" mean in this particular experiment? Not being specific causes your readers to stop and wonder. Using almost the same number of words you could write

With our cathode-ray tube, we measured fast electrons (10^7 meters/second) in the arc discharge.

Note that the word "fast" is important in this sentence; it qualifies the 10^7 meters/second measurement. Just saying

With our cathode-ray tube, we measured electron velocities at 10^7 meters/second in the arc discharge.

would not have been as precise because your readers might not know how 10^7 meters/second compares to other electron velocities.

Being specific does not mean that you eradicate general statements from your writing. General statements are important to give readers an overview of what will come or a summary of what has been said. Being specific means that you ground general statements; it means that you provide specific examples and details to give your general statements meaning. Writing that relies solely on general statements is empty. Consider this paragraph from a U.S. Office of Technology Assessment report:

Some options seem capable of developing more quickly than others for a given investment. Some promise costs which are lower than others. Some are better suited to small, remote applications; others to large centralized facilities. Some may only be used to generate electrical power; others can generate usable heat as well. Some may easily be modified to include thermal energy storage; others might not offer that choice. Some require large quantities of water, some do not. Some use more established, proven components; others are using riskier components hoping that the gamble eventually may prove to be a good one, and that the choice will give them a competitive edge.

Not one statement is specific. This paragraph presents readers with no answers, only needless questions. How much more quickly do some options develop? How much lower in cost are some options than others? You should anchor general words such as "some," "quickly," "lower," "small," "large," and "good" with specific numbers and examples. Otherwise, your writing floats—ungrounded.

AVOIDING OVERSPECIFICATION

In scientific writing, precision does not mean the highest degree of accuracy; it means the accuracy that is appropriate for your research and audience. Many times, scientists say too much. Consider this sentence from a progress report at a power plant:

> *Operations at the plant stopped momentarily because the thermal storage charging system desuperheater attemperator valve was replaced.*

Unless there had been special problems with that particular valve, a more appropriate level of accuracy for this situation would have been

> *Operations at the plant stopped for two hours so that a valve in the thermal storage system could be replaced.*

In this revision, the location of the valve was made less specific, while the time that the plant was down was made more specific. This report's principal reader happened to be the plant manager who cared more about how long the plant was down than about which particular valve was leaking.

Another example:

> *Within the master control system, there are four computers and approximately 983 microprocessors.*

Saying that there are "approximately 983 microprocessors" is silly. The engineer should either say

> *The master control has four computers and 983 microprocessors.*

or

> *The master control has four computers and about one thousand microprocessors.*

Packing sentences with too many details also makes for tiresome reading:

> *A 1-mm diameter, 656-nm beam with uniform intensity across the beam was produced by using a wavelength/polarizer combination to split off part of the 532-nm output from the Nd:YAG laser to pump a second dye laser (Laser-Ray LRL-2, also operated with DCM dye) with a side-pumped configuration for the final amplifier, and selecting the central portion of the collimated beam with an aperture.*

How could the physicist who wrote this sentence expect anyone to follow it? There are too many details here. Are all these details necessary? Couldn't the physicist have given these details in several sentences? Better still, couldn't the physicist have placed some of the secondary details (such as beam wavelength and the manufacturer's name) in a schematic? Scientists sometimes worry so much about telling readers everything that they end up not informing readers of anything.

Being precise doesn't mean compiling details; it means selecting details. You must choose details that inform.

> *The fuel pellets used in inertial confinement fusion are tiny, the size of BBs, but they are potentially the most powerful devices mankind has ever known. If we can compress the fuel in the pellets to a plasma, the fuel's deuterium and tritium atoms can overcome their mutual electrical repulsion and fuse into helium atoms, giving off energy ($E = mc^2$). The power needed to ignite fusion in the pellets is 100 trillion watts; however, the power released from the fusion is one hundred times that much.*

This paragraph informs; it informs because the physicist selected only the most important details about the pellets. Because scientific writing is compressed, you only have room for the most important details. Make them count.

CHAPTER 5

Being Clear

When you are out to describe the truth, leave elegance to the tailor.

Albert Einstein

While precision in scientific writing means choosing the right words, clarity means not choosing any wrong ones. Too often in scientific writing, an ambiguous phrase or sentence disrupts the continuity—and authority—of an entire section of precise writing. In scientific writing, each sentence builds on the ones around it. If one sentence is weak, your language falters and your readers stumble.

KEEPING IT SIMPLE

If there was only one piece of stylistic advice I could whisper in the ear of every scientist, it would be to *keep it simple*. Undoubtedly, the most frequent language mistake in scientific writing is needlessly complex prose.

Keeping your phrases simple. Don't string adjectives in front of your nouns. Nouns, particularly subject nouns, are anchors in sentences, and stringing adjectives in front of them tires readers. Readers lose patience waiting for the anchor to break water.

The junction is a right angle gusset reinforced butt joint.

Imagine the engineer who wrote this sentence actually saying it out loud. She would lose her breath before finishing. If all these adjectives are important, then the engineer should either work them into two sentences or else place them in phrases and clauses around the noun.

The junction is a butt joint that is right-angled and gusset reinforced.

Stringing adjectives dilutes the adjectives' meanings. The adjectives become lost in the string. Moreover, long strings bring imprecision into sentences:

The decision will be based on economical fluid replenishment cost performance.

What exactly does the phrase "economical fluid replenishment cost performance" mean? Will the decision be based on the performance of the fluid or on the cost of replacing the fluid or on something else? Pursuing our goals of precision and simplicity, we revise this sentence to

We will base the decision on the cost of replacing the thermal oil.

Keeping your sentences simple. In scientific writing, most sentences are too complex. They average over twenty-five words, which is long. The average sentence length in *Newsweek* is only about seventeen words. More important, many sentence structures in scientific writing are convoluted.

The object of the work was to confirm the nature of electrical breakdown of nitrogen in uniform fields at relatively high pressures and inter-electrode gaps which approach those obtained in engineering practice, prior to the determination of the processes which set the criterion for breakdown in the above-mentioned gases and mixtures in uniform and nonuniform fields of engineering significance.

This sentence is complex. It is needlessly complex. It has sixty-one words, two of which are hyphenated. Also, the sentence structure is convoluted. After the word "gaps," the sentence wanders aimlessly from one prepositional phrase to another. Prepositional phrases are important; they provide transition for incorporating details. This sentence, however, has eleven prepositional phrases; count them—eleven. Prepositional phrases act like boxcars on a train. They provide no momentum, only friction. Something else that hurts this sentence is that it has no symmetry—either structurally or logically. It just wanders.

Would the sentence be strengthened by breaking it up into shorter sentences? Yes, but don't take shorter sentences as a panacea for unclear writing. Although you could strengthen most scientific papers by making all the sentences short, using only short sentences will not produce strong writing. You need long and medium-length sentences to keep your writing from sounding choppy as well as to provide variety and emphasis. Moreover, it is not long sentences that confuse readers; it is complex sentences that confuse readers. Consider the same information presented with our rules of simplicity and precision:

> *At relatively high pressures (1 atm) and typical electrode gap distances (1 mm), we studied the electrical breakdown of air in both uniform and nonuniform fields.*

Twenty-six words, a fairly long sentence. Nevertheless, the sentence is strong; it is precise and clear.

So now you ask, "How long should a sentence be?" There is no easy answer. Depending on the situation, a sentence can be two words or sixty. Perhaps a better question is, "When is a sentence too long?" A sentence is too long when it confuses your readers.

> *To separate the hot and cold oil, one tank was used that took advantage of the thermocline principle which uses the rock and sand bed and the variation of oil density with temperature (8% decrease in density over the range of operating temperatures) to overcome natural convection between the hot and cold regions.*

Does this sentence inform? No. This sentence is indigestible; there is too much to chew. The writer has presented many details, but these details are lumped like unstirred oatmeal.

Before fixing this sentence, you need to decide which details in the sentence are important. Does the reader really need to know how the hot and cold oil are separated or how the density varies in the tank or what the thermocline principle is? If these details are not important, then you should delete them. However, if the reader needs to know these details to understand the research, then you should present them in digestible portions:

> *The question was how to separate the hot and cold oil in the rock and sand bed. Rather than have one tank hold hot oil and another tank hold cold oil, we used a single tank for both. This design took advantage of the variation of oil density with temperature. In our storage system, oil decreases in density by 8% over the range of operating temperatures. This variation in density allows the hot oil to float over the cold oil in the same tank. Natural convection is impeded by the position of hot over cold and by the rock and sand bed. Thus, the heat transfer between hot and cold regions is small because it occurs largely by conduction. The concept of storing heat in a single vessel with hot floating over cold is known as "thermocline storage."*

Many scientists hold the misconception that scientific writing should be written in the style of the seventeenth century. These scientists are loyal to their mentors; they adopt their mentors' philosophies, their mentors' idiosyncrasies in the lab, and—unfortunately—their mentors' writing styles. Where did their mentors' writing styles come from? From their own mentors. Compared to the styles of other genres such as fiction or journalism, the styles of scientific writing change little from generation to generation. Too many scientists, great ones included, write in the formal complexity of the seventeenth century. Consider an excerpt from Niels Bohr's writing:

> ***The Correspondence Principle.*** *So far as the principles of the quantum theory are concerned, the point which has been emphasized hitherto is the radical departure from our usual conceptions of mechanical and electrodynamical phenomena. As I have attempted to show in recent years, it appears possible, however, to adopt a point of view which suggests that the quantum theory may, nevertheless, be regarded as a rational generalization of ordinary conceptions. As may be seen from the postulates of the quantum theory, and particularly the frequency relation, a direct connection between the spectra and the motion of the kind required by the classical dynamics is excluded, but at the same time the form of these postulates leads us to another relation of a remarkable nature . . .*[1]

This writing is needlessly complex. There are many bloated phrases such as "which has been emphasized hitherto." The sentences are also long. They average almost forty words; but more important, their structures are convoluted.

"Ah," you say, "but this writing is elegant and beautiful."

Is it? The ideas are beautiful, but the writing is murky, almost inaccessible. In scientific writing, beauty lies in clarity and simplicity.

> ***The Correspondence Principle.*** *Many people have stated that the quantum theory is a radical departure from classical mechanics and electrodynamics. Nevertheless, the quantum theory may be regarded as nothing more than a rational extension of classical concepts. Although there is no direct connection between quantum theory postulates and classical dynamics, the form of the quantum theory's postulates, particularly the frequency relation, leads us to another kind of connection, one that is remarkable . . .*

When writing, you should imagine yourself sitting across from your most important reader. Write your paper as if you were talking to that reader. This doesn't mean that your writing should be informal. Rather, it means that you should rid your writing of needless formality. The purpose of scientific writing is to inform, not to impress. Therefore, don't write

> *In that the "Big Bang," currently the most credible theory about how the universe was created, explains only the creation of hydrogen and helium, we are left to theorize as to how all the other elements came into being. Having studied the nuclear reactions that constitute the life and death cycle of stars, many scientists believe that therein lies the key.*

This kind of writing bewilders. It is not for twentieth century readers. If you need a writer to emulate, choose Einstein or Feynman. They would have written the above paragraph more like this:

> *The "Big Bang" is the most credible theory for the creation of the universe. Nevertheless, it only explains the creation of helium and hydrogen. What about the other elements? Many scientists think that they came from nuclear reactions in the life and death cycles of stars.*

AVOIDING VAGUENESS

Avoiding vagueness is parallel advice to being specific. Sometimes in writing, though, you must edit from both sides of the coin. On one edit, you must worry about what to put in; then on another edit, worry about what to take out.

One sure way to avoid vagueness in your writing is to remove abstract nouns. Many nouns such as "factor," "condition," and "parameter" have little or no meaning; they are generic. Not only do they waste reading time, they also attract superfluous words that cloud important details. Therefore, don't write

The nature of this problem, which had many parameters, led us to the conclusion that we needed a computer approach with a Cray computer.

This sentence is needlessly vague. Revision gives

Because the equation had forty-five variables, we used a Cray computer.

Here is a short list of abstract nouns that continually creep into scientific writing:

ability	factor
capability	nature
concept	parameter

There are others. Many times, these abstract nouns precede the word "of." Removing these words will strengthen your writing.

ELIMINATING OVERSIGHTS

Many ambiguities in scientific writing are difficult to classify.

The solar collector worked well under passing clouds.

Does the solar collector work at a height that is well below the height of passing clouds, or under passing clouds, does the solar collector work well? Your English teacher would probably say that this ambiguity is a careless writing mistake. Perhaps it is, but it's not so much a careless writing mistake as a careless editing mistake. Everyone makes writing mistakes on early drafts. Strong writing comes from working hard on your revisions, not from conjuring magic on your first drafts.

Consider another editing oversight; this one from a paper presenting the design of a system for measuring solar radiation:

> *Two general requirements to be met are (1) to survive and accurately measure the radiation incident on the receiver and (2) to present the data in a form which can be used to verify computer code predictions.*

The engineer did not smooth the "trail" here. The sentence has gaps. Who or what must survive? Moreover, the engineer says that there are two requirements, yet lists three. Attention to precision and clarity gives this revision:

> *The radiometer system must meet three requirements: (1) it must accurately measure (to within 5%) the solar radiation on the receiver; (2) its electronics must survive in solar radiation as intense as 300 kilowatts per square meter; and (3) its output must be able to verify computer codes.*

Note that there are many ways to rewrite the sentence. The point is that the sentence needs rewriting. You do not attain clarity on first drafts; you attain clarity as you polish drafts and have all your ideas on paper in front of you.

USING PRONOUNS

A pronoun refers to the last noun used. Although most good writers stretch this rule somewhat, they are careful to check for ambiguities. Many scientists, however, ignore this rule. They abuse pronouns, particularly the pronouns "it" and "this."

> *To protect the radiometer from overheating in the high flux environment of the receiver, it was mounted in a silver-plated stainless steel container.*

What is mounted in the container? The radiometer? The receiver? The environment? The rules of grammar say that the "it" refers to the receiver. Moreover, the sentence reads as if the "it" refers to the receiver, and in this particular report the readers had no reason to think the "it" was not the receiver until twenty pages later when the engineer stated that the receiver was over forty-five feet tall. The engineer had intended the "it" to refer to the radiometer, although the radiometer was two nouns removed.

The way many scientists treat the pronoun "it" is unsettling, but the way many scientists treat the word "this" is criminal:

There are no peaks in the olefinic region (near 5.5 ppm). Therefore, no significant concentration of olefinic hydrocarbons exists in fresh oil. This places an upper limit on the concentration of olefins—no more than 0.01%.

What does the chemist want the "this" to refer to? To the last noun of the previous sentence—oil? To the subject of the previous sentence—concentration? To the idea of the previous sentence—that there is no significant concentration of olefinic hydrocarbons? Actually, the "this" refers to none of these. The chemist intended the "this" to refer to the lack of peaks in the olefinic region.

The word "this" is a directive, a pointer. To use "this" in a sentence as an anchor (as a subject noun) is weak writing. Instead, let the word "this" point.

The chromatogram has no peaks in the olefinic region (near 5.5 ppm). Therefore, no significant concentration of olefinic hydrocarbons exists in fresh oil. This chromatogram finding places an upper limit on the olefin concentration, no more than 0.01%.

USING EQUATIONS

Often in scientific writing, the most efficient way to convey relationships is through mathematical and chemical equations. Without equations, even simple relationships can appear complex:

The absorptance is calculated as one minus the correction factor times the measured reflectance.

This scientist made his readers work too hard. An equation would have communicated the relationship much more efficiently.

The absorptance (A) is calculated by

$$A = 1 - kR,$$

where k is the correction factor and R is the measured reflectance.

Although equations simplify relationships, they still make for difficult reading. Readers must stop and work through the meaning of each variable. Therefore, anytime you introduce equations into your writing, you should show why those equations are important. You should give readers incentives to push through the work of understanding your equations.

The reaction $O_2 + H \rightarrow O + OH$ is the single most important chemical reaction in combustion; it is responsible for the chain-branching of all flame oxidation processes.

Because equations are difficult to read, you want to make them as clear as possible. One thing you should do is define all the terms in your equations. Don't assume that all your readers will read **n** as the gas phase reaction order or $\boldsymbol{\sigma}$ as the ratio of unburned to burned temperatures.

The burning rate (Y) of a homogeneous solid propellant is given by

$$Y = \frac{\rho}{\alpha} (2\lambda\hat{\tau})^{1/2} \, \mathrm{Le}^{n/2} \left[\frac{c(1 - \sigma)}{(c - \sigma)(1 + \gamma_s)} \right],$$

where ρ, c, and λ are the gas-to-solid ratios for density, heat capacity, and thermal diffusion, $\hat{\tau}$ is a scaled ratio of the rate coefficient for the gas-phase reaction, γ_s is the ratio of the heat of sublimation to the overall heat of reaction, Le is the gas phase Lewis number, and α is the fraction of propellant which sublimes and burns in the gas phase.[2]

Besides stating the importance of equations and defining all variables, you should also clearly state your equations' assumptions.

The Townsend criterion for the electrical breakdown of a gas is given by

$$\frac{\omega}{\alpha}\left(e^{(\alpha d)} - 1\right) = 1,$$

where α is the primary Townsend ionization coefficient, ω is the secondary Townsend ionization coefficient, and d is the distance between electrodes. This criterion for breakdown is actually a physical interpretation of conditions in the gas rather than a mathematical equation because Townsend derived his theory for a steady state case that has no provisions for transient events such as breakdown.

Without the caution in the last sentence, someone could misunderstand Townsend's criterion. Readers need to know exactly when your equations apply and when they don't.

When presenting equations, you should also use examples to give readers a feel for the numbers involved.

In the first stage (Stage I) of an electron avalanche, diffusion processes determine the avalanche's radial dimensions. The avalanche radius r_d is given by

$$r_d = (6Dt)^{1/2},$$

where D is the electron diffusion coefficent and t is time. In the regime of interest (where voltages are greater than 20% of self-breakdown), the time of development is so short that little expansion occurs. For the case of nitrogen at atmospheric pressure, $D \simeq 862\frac{cm^2}{sec}$, $t \sim 10^{-8}$sec, and $r_d \sim 7.2 \times 10^{-3}$cm.

As the number of electrons in the avalanche increases, electrostatic repulsion begins to play a role in the expansion of the avalanche. In this stage (Stage II), the avalanche increases exponentially with time. Taking space charge into account, we express the avalanche radius by

$$r_e = \left(\frac{3e}{4\pi e_o E_o}\right)^{\frac{1}{3}} e^{\frac{\alpha z}{3}},$$

where e is the electron charge, E_o is the external electric field, α is the Townsend primary ionization coefficient, ϵ_o is the permittivity of free space, and z is the average position along the gap axis of the avalanche with z = 0 being the cathode position. For breakdown in nitrogen at atmospheric pressure and *an applied field of* $40\frac{kV}{cm}$ *(which corresponds to a voltage ~25% above self-breakdown), the avalanche radius* r_e *is 0.14 cm at z = 0.05 cm.*[3]

Without these two examples, many readers would not realize the huge size difference between the avalanche radii of Stage I and Stage II. Examples anchor your equations with details your readers will remember.

Although you might have shown an equation's importance, identified its symbols and assumptions, and even given examples, you still may not have done enough. For effective writing, you must convey the physical meaning of the equation. As Dirac said, "I understand an equation if I have a way of figuring out the characteristics of its solution without actually solving it." That is the type of understanding you want to give your readers.

In 1924, Louis De Broglie made an astute hypothesis. He proposed that, because radiation sometimes acted as particles, matter should sometimes act as waves. De Broglie did not base his hypothesis on experimental evidence; instead, he relied on intuition. He believed that the universe was symmetric. Using Einstein's energy relation as well as the relativistic equation relating energy and momentum, De Broglie derived an equation for the wavelength of matter, λ_m*:*

$$\lambda_m = \frac{h}{p_m} .$$

In this equation, h is Planck's constant and p_m *is the matter's momentum. In 1927, Davisson and Germer experimentally verified De Broglie's hypothesis . . .*

This discussion could have introduced De Broglie's equation with a detailed mathematical derivation. Instead, the discussion presented De Broglie's argument for symmetry. A rigorous mathematical derivation is not always the best way to communicate an equation.

PUNCTUATING CORRECTLY

Punctuation rules are important. They are devised to eliminate ambiguities in language. Learn punctuation. Find a handbook such as *The Chicago Manual of Style* and keep it on your writing desk. Few things undercut the authority of a piece of writing more than a simple mistake in punctuation.

The Period

In scientific writing, periods aren't used enough to end sentences. Too many sentences go on and on, taxing the reader's concentration. Although you should have a variety of long, short, and medium-length sentences, You should hold your average sentence length to no more than seventeen words, not twenty-seven words as in so many journal articles.

Although you should generously use periods to end sentences, you should avoid using periods to abbreviate. When used in abbreviations, periods often trip readers; readers think they've come to the end of the sentence.

> *Fig. 1.1 shows a gamma-ray line, i.e. radiation at a single gamma-ray energy level, which theorists had predicted would result from N. Cygni.*

This sentence is choppy. By varying the punctuation and cutting needless abbreviation, you can make a much smoother sentence:

> *Figure 1-1 shows a* gamma-ray line *(radiation at a single gamma ray energy level) that theorists had predicted would result from Nova Cygni.*

The Comma

Commas give scientists headaches. Some scientists worry about having too few commas. These scientists paint their sentences with commas; they use commas anywhere that there's the slightest suggestion of a pause, thus making readers wade through each sentence.

> *Although acquired immunodeficiency syndrome, also known as AIDS, does not transmit easily, we think that, within the next few years, it will spread to every sect of society.*

This sentence reads too slowly. You should cut the commas after "that" and "years."

Other scientists scorn commas. These scientists will use

commas only in the most extreme cases, and sometimes not even then, thus making readers trip over ambiguities.

After cooling the exhaust gases continue to expand until the density which was high to begin with reaches that of free-stream.

This sentence needs a comma after "cooling" and a set of commas around the clause "which was high to begin with."

There are many rules for commas. Some rules are unwavering. For instance, in a series of three or more items you should use commas to separate each term. Thus, write

hydrogen, oxygen, and nitrogen.

You should also use commas to set off contrasted elements (these expressions often begin with *but* or *not*).

The shark repellent using 20% copper acetate and 80% nigrosene dye was quite effective with Atlantic sharks, but ineffective with Pacific sharks.

Many reported injuries result from shark bumps, not shark bites.

Although these rules for commas in a series and commas setting off contrasting elements are unwavering, many comma rules are hazy. For example, using a comma after an introductory phrase depends on the situation.

In many cases the infected people developed antibodies to the AIDS virus and remained healthy.

Placing a comma after "cases" is pretty much optional. Few readers would notice if you did or didn't. Some introductory phrases, however, require a comma.

When feeding a shark often mistakes undesirable food items for something it really desires.

You need a comma after "feeding."

Is there any practical way to approach using commas? Well, first you should realize that the purpose of commas is to eliminate ambiguities. Comma rules aren't arbitrary. They have a specific purpose: to prevent readers from tripping over language. Therefore, when unsure about a comma, think about whether your readers would trip if the comma weren't there. If your readers would not trip, then remove the comma. Be consistent in your use of commas. If you punctuate a particular sentence structure one way in the beginning of a paper, then punctuate it the same way throughout.

The Colon

Colons introduce lists.

We studied five types of marsupialia: opossums, bandicoots, koalas, wombats, and kangaroos.

Colons should not, however, break continuing statements.

Incorrect:

The five types of marsupialia we studied were: opossums, bandicoots, koalas, wombats, and kangaroos.

Correct:

The five types of marsupialia we studied were opossums, bandicoots, koalas, wombats, and kangaroos.

Besides introducing lists, colons are also used for definitions.

The laboratory growth of this germanium crystal made possible a new astronomy tool: a gamma-ray detector with high-energy resolution.

The phrase after the colon defines the "new astronomy tool."

The Semicolon

In scientific writing, the semicolon is misunderstood. Scientists toss semicolons into sentences whenever they're unsure what punctuation to use. The semicolon is optional—you don't have to use it. In fact, many good writers never do. The semicolon has a specific purpose: to connect two sentences closely linked in thought.

There is no cure for Alzheimer's disease; it brings dementia and slow death to thousands of Americans every year.

The Dash

The dash sets off parenthetical remarks.

The unique feature of their design is a continuous manifold, which follows a unidirectional—as opposed to serpentine—flow for the working fluid.

Dashes are also used for phrases and clauses that would cause ambiguities if set off by commas.

After one year, we measured mirror reflectivity at 96%—a high percentage, but not as high as originally expected.

Be careful with the dash. Too many dashes will break the continuity of your writing.

REFERENCES

1. Niels Bohr, *The Theory of Spectra and Atomic Constitution,* (Cambridge: At the University Press, 1924), p. 81.
2. S. B. Margolis and R. C. Armstrong, "Two Asymptotic Models for Solid Propellant Combustion," *Sandia Combustion Research Program Annual Report,* (Livermore, CA: Sandia National Laboratories, 1985), Chap. 5, pp. 6–8.
3. E. E. Kunhardt and W. W. Byszewski, "Development of Overvoltage Breakdown at High Pressure," *Physical Review A,* 21, no. 6 (1980), pp. 2069–72.

CHAPTER 6

Being Familiar

The whole of science is nothing more than a refinement of everyday thinking.

Albert Einstein

To inform your audience, you must use language that your audience understands. It is often difficult to find language that is familiar to your audience. Most scientific research is specific. Scientists work on particular crooks of particular ridges of particular mountains. Every mountain has its own particular set of words and abbreviations.

Plasma Physics:	solitons	MHD
Hemodynamics:	atherosclerosis	GlyHgb
Spectroscopy:	photoisomerization	HgD
Solar Energy:	heliostat	CRS

In a scientific paper, the writer—not the reader—bears the responsibility of bridging the language gap. When you write a scientific paper, you must constantly ask yourself whether your word choices are familiar. If a sentence calls for the word "heliostat" and if "heliostat" is the most precise word for that sentence, then you must use it. If you think your audience is unfamiliar with "heliostat," then you must define it or provide an analogy to explain it.

DEFINING UNFAMILIAR WORDS

One of the most important considerations you can give your audience is to define unfamiliar words. Which words you define depends on your audience. When writing for an audience outside your research mountain, you should define any unusual word particular to your research mountain—perhaps a word such as "bremsstrahlung." You should also define any common word such as "receiver" that assumes a particular meaning on your research mountain.

What is the best way to define a word? If a definition is short, you can include it within the sentence that the word is used:

> *In a solar central receiver system, a field of sun-tracking mirrors or* heliostats *focuses reflected sunlight onto a solar-paneled* boiler *or* receiver *mounted on top of a tall tower.*

> *The thermal oil is not a pure compound but a mixture of many* aliphatic hydrocarbons *(paraffins).*

When the definition is complex or unusual, you should expand your definition to a sentence or two.

> Bremsstrahlung *(from the German:* bremse *(brake), and* strahlung *(radiation)) is the radiation emitted by a charged particle that is accelerated in the Coulomb force field of a nucleus.*

> Atherosclerosis *is a disease in which fatty substances line the inner walls of arteries. If these deposits plug an artery, a heart attack, stroke, or gangrene may occur.*

In your definitions, select words that your readers are familiar with; otherwise, your definitions won't inform.

> Serum Lipids *are cholesterols and triglycerides transported in the blood.*

If your readers do not know the meaning of "serum lipids," chances are they don't know the meaning of "triglycerides" either. You should either replace "triglycerides" with a familiar word, or else define "triglycerides" within your definition of "serum lipids."

When formally defining a term, you should begin the definition with a familiar noun that identifies the class to which the term belongs. You should then provide enough information to separate that term from all other terms in the class. Don't write

Cholesterol *is present in body fluids and animal cells.*

or

Triglycerides *are fats.*

As formal definitions, these definitions are incomplete. The first definition needs a noun such as "fat" to classify "cholesterol," and the second definition needs additional information to separate "triglycerides" from all other kinds of fats.

You should show some constraint when using definitions. Too many definitions will break the flow of your paper. If you have several definitions in a report, you should include a glossary.

AVOIDING JARGON

Jargon is vocabulary particular to a laboratory or company. Jargon may be abbreviations or slang terms. Don't assume that jargon is inherently bad. For communication *within* a large laboratory, jargon can concisely identify experiments and buildings. Let's say you work at a laboratory called the Pulsed Power Facility. When you write internal memos and reports, you might use the facility's jargon:

PPF—*Pulsed Power Facility*

Jarva—*a neodymium-glass laser at the Pulsed Power Facility*

For readers who work at the facility, this jargon poses no problem. However, when you write an external report or journal paper, this jargon alienates readers. If the jargon is slang, such as "Jarva," write around it. Just use "the neodymium-glass laser" the first time, and "the laser" thereafter. In a pulsed power paper, what difference does it make to your audience whether your laser is named "Jarva" or "Olivia"?

If the jargon is an abbreviation, such as "PPF," and occurs only a couple of times, then you should avoid the abbreviation and write out the expression: "Pulsed Power Facility." If the abbreviation occurs several times, then you should define the abbreviation the first time it's used:

> *Acquired immunodeficiency syndrome, also known as AIDS, is a disease that breaks down part of a body's immune system. What causes AIDS? AIDS is caused by a virus.*

"AIDS" happens to be an unusual abbreviation in that it is more familiar than the expression it abbreviates.

For strong scientific writing, you should keep the number of unfamiliar abbreviations to a minimum. Many scientists unfortunately go out of their way to include every unusual acronym and abbreviation from their research.

> *For the first year, the links with SDPC and the HAC were not connected, and all required OCS input data were artificially loaded. Thus, CATCH22 and MERWIN were not available.*

This paragraph reads like a cryptogram. Whenever precision allows, you should use common words, not unfamiliar abbreviations.

> *Because some of the links in the computer system were not connected the first year, we could not run all the software codes.*

Jargon not only alienates; it often misleads.

> *The 17% efficiency of water/steam system was a* showstopper.

Someone not familiar with the expression "showstopper" might interpret 17% as a favorable efficiency, so favorable that the show was interrupted by applause. In this particular sentence, however, the engineer meant just the opposite: 17% was so poor that the system was disregarded and the research was stopped.

USING EXAMPLES

One of the most effective ways to inform is through example. Scientists know the value of examples; the best mathematics textbooks were always the ones with many sample problems. Whenever you make general statements, you should anchor them with examples. Don't leave your reader clutching to generalities.

General Writing:

Since the design of our solar power plant, significant advances have been made in solar energy technology.

Specific Writing:

Since the design of our solar power plant, significant advances have been made in solar energy technology. For example, experimental tests have shown that using molten salt (rather than water) as the heat transfer fluid could increase overall system efficiency by as much as 8%.

The first statement will be soon forgotten because it relies solely on the phrase "significant advances." The revision, however, anchors the phrase with a specific example that readers will remember.

USING NUMERICAL ANALOGIES

Scientific writing is full of numerical findings. Many times, though, readers need analogies and comparisons to understand the significance of those numbers.

In the brightness tests, the maximum retinal irradiance was less than 0.016 W/cm^2, a brightness about one-fourth that of a household light bulb.

During daylight hours, the Solar One Power Plant produced 10 megawatts of electric power, enough electricity to service about 3500 homes.

Numerical analogies make your writing unique. Consider how Feynman gives significance to the magnitude of electrical forces.

If you were standing at arm's length from someone and you had one percent more electrons than protons, the repelling force would be incredible. How great? Enough to lift the Empire State Building? No! To lift Mount Everest? No! The repulsion would be enough to lift a weight equal to that of the entire earth.[1]

USING NUMERALS

In their writing, scientists overuse numerals. Numerals are actual figures: 0, −1, 2.76, 3000. Whenever possible, you should avoid using numerals in your text because numerals slow the reading. When numbers can be expressed in two words or less, write them out.

one

four thousand

ten million

seventy-six

There are exceptions to this rule:

negative numbers	−1
numbers with decimals	0.2
specific measurements	12 meters/second
page numbers	page 56
illustration numbers	Figure 13
percentages	15%
monetary figures	$200,000

Also, don't begin a sentence with a numeral. If a numeral is called for, then restructure the sentence so the numeral doesn't appear first.

Incorrect:

. . . measured corrosion. 64.1 milligrams of copper corroded during the tests.

Correct:

. . . measured corrosion. During the tests, 64.1 grams of copper corroded.

REFERENCES

1. Richard P. Feynman, *The Feynman Lectures on Physics*, (Reading, MA: Addison-Wesley Publishing Co., 1964), II, 1.

CHAPTER 7

Being Forthright

Short words are the best and old words when short are the best of all.

Winston Churchill

The purpose of scientific writing is to inform. Therefore, the language of scientific writing should be forthright; it should be sincere and straightforward.

USING STRONG NOUNS

Nouns provide anchors in sentences. Don't bury them beneath a sea of adjectives.

The existing nature of Mount St. Helens' volcanic ash spewage was handled through the applied use of computer modeling capabilities.

Adjectives hinder readers from finding the anchors. You should choose specific nouns that need few adjectives.

With computers, we modeled how much ash spewed from Mount St. Helens.

USING STRONG VERBS

Verbs provide momentum in sentences. When your verbs are strong, your writing moves. Many scientists sap strength from their verbs by burying them as nouns in weak verb phrases:

Weak Verb Phrase	Strong Verb
performed the development of	developed
made the arrangement for	arranged
made the measurement of	measured
made the decision	decided

Weak verb phrases cripple your sentences.

This year's use of the storage system is not only for the provision of sealing steam to the boilers, but also for the production of electricity for the grid.

Forthright writing frees verbs from weak verb phrases.

This year, the storage system not only provides sealing steam to the boilers, but also produces electricity for the grid.

The first sentence just sits. The second sentence moves. It moves because the verbs are active. "Provides" and "produces" have replaced the passive verb "is."

The verb "to be" is passive. Many times, you must use "to be" verbs, such as when you are defining or equating terms.

A positron is a positively charged electron produced in the beta decay of neutron-deficient nuclides.

Many scientists, however, overload their writing with unnatural "to be" verbs. These needlessly passive verbs slow the reading. When editing your writing, you should question all "to be" verbs, especially when they appear near specific verbs:

Needlessly Passive	Active
is used to detect	detects
is beginning	begins
is shadowing	shadows

Needlessly Passive	**Active**
is following	follows
is capable of	can

Often, you can strengthen the language by rearranging sentences to replace "to be" verbs with more active verbs.

Passive Writing:

Sputter Ion Mass Spectroscopy was used *to determine the position of isotopically tagged atoms. The sample gas* was mounted *in a vacuum chamber and an ion beam* was used *to sputter atoms from the surface. A fraction of the sputtered atoms* was ionized *and* was detected *using a quadrupole mass analyzer.*

Active Writing:

Sputter Ion Mass Spectroscopy determined *the position of isotopically tagged atoms. An ion beam* sputtered *atoms from the surface of a sample mounted in a vacuum chamber. Then, a quadrupole mass analyzer* detected *the ionized fraction of sputtered atoms.*

Many scientists hold the misconception that scientific papers and reports should be written in the passive voice. However, the purpose of scientific writing is to inform as efficiently as possible, and the most efficient way to inform is through strong, straightforward writing—writing that uses the active voice. Needlessly passive verbs slow your writing; they reduce your writing's efficiency.

The feedthrough was composed *of a sapphire optical fiber, which* was pressed *against the pyrotechnic that* was used to confine *the charge.*

Eliminating the passive voice from this sentence strengthens the writing:

The feedthrough contained *a sapphire optical fiber, which* pressed *against the pyrotechnic that* confined *the charge.*

Don't go overboard and assume that all passive voice is wrong. Many times, you must use the passive voice, such as when the subject of your writing is acted upon:

> *On the second day of our wildebeest study, one of the calves wandered just a few yards from the herd and* was attacked by *wild dogs.*

In this example, there is nothing wrong with the passive verb "was attacked." In fact, eliminating passive voice from this particular sentence makes the writing awkward:

> *On the second day of our wildebeest study, one of the calves wandered just a few yards from the herd and wild dogs attacked the calf.*

Much passive voice arises in scientific writing because scientists cling to the misconception that they can't use the first person ("I" or "we"). Well, Einstein used the first person. He was not only a great scientist, but a great scientific writer. Faraday used the first person; so did Watson, Crick, Curie, Darwin, Lyell, Schockley, Feynman, Freud, and Lister. The second constraint of scientific writing says that you must remain honest. Using the first person doesn't keep you from that honesty. As long as the emphasis of your writing remains on your research and not on you, there is nothing wrong with using the first person in your papers.

Also, avoiding the first person leads to unnatural wording:

> *In this paper,* the author assumed that *all collisions were elastic.*

The phrase "the author assumed that" is silly. The reader can see the author's name on the paper. Using the word "author" instead of "I" or "we" suggests that the writer of the paper wasn't the researcher. Other awkward phrases include

> It was determined that . . .
>
> It was decided that . . .
>
> It was realized then that . . .

These phrases suggest that there was some absolute force—the IT force—writing the paper. This suggestion does break the second constraint of scientific writing, namely that you stay honest. The IT force suggests that your conjectures are more than conjectures, that some being greater than mortal man performed the research.

Perhaps the strongest reason for scientists to use the first person is that it reminds them that they're responsible for what is written and what will be read.

WRITING ANALYTICALLY

In any description, a writer must present details. There are two ways to present details in a sentence: catalogical writing and analytical writing.[1] In catalogical writing the writer does not give emphasis to any details; therefore, minor details appear as important as major details.

One of the panels on the north side of the receiver will be repainted with Solarcept during the February plant outage.

What is the most important detail in this sentence? Is it that the panel is on the north side? Is it that the panel is being repainted with Solarcept? Is it that the repainting will occur during the February plant outage? The problem with this sentence is that you don't know. All details have the same weight. In catalogical writing, prepositional phrases are overused. As was stated in the chapter on clarity, prepositional phrases provide no momentum to sentences, only friction. Moreover, they give no emphasis to details. Details when linked by prepositional phrases have the same amount of importance.

In analytical writing, however, weights are assigned to details so that important details stand out.

Because the February plant outage gave us time to repair the north side of the receiver, we repainted the panels with Solarcept, a new paint developed to increase absorptivity.

In this sentence, reasons are given for the details. The reader sees why things happened. In analytical writing, the writer anticipates the readers' questions and answers them. Where catalogical writing overuses prepositional phrases, analytical writing judiciously uses *dependent clauses* and *infinitive phrases*. Dependent clauses are clauses that begin with introductory words such as "because," "since," "as," "although," and "when."

Because the February plant outage gave us time to repair the north side of the receiver, . . .

Infinitive phrases are verb phrases that begin with "to" and give reasons why things happened.

. . . to repair the north side of the receiver.

. . . to increase absorptivity.

AVOIDING PRETENTIOUS WORDS

Many times, to be precise in scientific writing, you must use unfamiliar words. Words such as "polystyrene," "glycosuria," and "soliton" have no simple subsitutes, but many words used in scientific writing are needlessly unfamiliar. Words such as "utilize," "implement," and "fabricability" are pretentious; they offer no precision, no clarity, and no continuity to writing. Moreover, they smack of a pseudo-intellectuality that poses a barrier between writer and reader. Some of these words—"interface," for example—are precise when used in certain contexts, yet imprecise and pretentious in others. Below is a short list of common pretentious words in scientific writing:

Approximate: a pretentious way to say "about."

Capability: a clue that you need to rewrite your sentence with an active verb preceded by the word "can." You should challenge all *-ability* words.

Implement: a farm tool. Many scientists, however, use "implement" as a verb. What they really want to write is "put into effect" or "carry out"—these verb phrases are old and simple. They're the verb phrases Winston Churchill would have used.

Interface: the interstitial boundary between two molecular planes. When used in this denotation, "interface" is precise. But many scientists use "interface" instead of the verb "to meet," or instead of the noun "junction." These abuses often produce awkward images:

In the development of the Five-Year Plan, Division 8475 hopes to interface with Division 8265.

The connotation of two groups of people actually interfacing is rather perverse, isn't it?

Networking: an example of a simple noun (network) convoluted into an awkward verb. There is no excuse for this kind of writing.

Utilize: a pretentious way to say the verb "use." You should challenge all *-ize* words.

Utilization: a pretentious way to say the noun "use." Scientists should challenge all *-ization* words. These words are clues that you need to rewrite the sentences with specific nouns and active verbs.

Here are some other pretentious words and their simple equivalents:

activate	start
component	part
consequently	so
contiguous	adjacent
demonstrate	show
discretized	discrete
facilitate	cause; make
finalize	finish
initialize	begin
prioritize	assess; assign priorities
prioritization	assessment
subsequently	then

Many of these words are long. Most of these words have French and Latin roots (rather than Anglo-Saxon). All of these words have infested scientific writing.

> *The goal of this study is to develop an effective commercialization strategy for solar energy systems by analyzing the factors that are impeding early commercial projects and by identifying the potential government and industry actions that can facilitate the viability of the projects.*

This sentence is inaccessible to readers. Revision with attention to precision, clarity, and forthrightness gives

This study will consider why current solar energy systems have not yet reached the commercial stage and will find out what steps industry and government can take to make these systems commercial.

AVOIDING ARROGANCE

In scientific writing, *tone* is whatever shows the writer's attitude toward his or her research. Because the purpose of scientific writing is to inform, not to impress or confuse, the tone of scientific writing should be forthright; it should be sincere and straightforward. The tone of scientific writing should not be arrogant. Many scientists, however, convey arrogance in their writing.

As is well known, the use of gaseous insulation is becoming increasingly more widespread, with gases such as air and sulphur hexafluoride featuring prominently. There has been some discussion of using gas mixtures like nitrogen and sulphur hexafluoride, and of course nitrogen is the major constituent of air . . .

These two sentences have problems with tone. The phrase "as is well known" is not forthright. If the readers know (and know well) the particular detail about gaseous insulation, then why did the writer mention it? If, however, the readers do not know the detail (as is probably the case), then the writer has assumed a superior position over the reader. The phrase "of course" in the second sentence strikes the same chord.

Another arrogant phrase often found in scientific writing is "clearly demonstrates."

The figure clearly demonstrates the ability of Raman spectroscopy to provide unambiguous chemical compound identifications from oxides as they grow on a metal surface.

When someone uses "clearly demonstrates" to describe a figure, more often than not the figure doesn't "clearly demonstrate" anything at all. Thus, readers are left wondering if the writer was pulling on sheepskin. The word "unambiguous" in the same example also reeks with arrogance; it defies readers to question the figure. A forthright revision would be

The figure shows that Raman spectroscopy can identify chemical compounds from oxides as they grow on a metal surface.

This revision is simple and straightforward. Research should stand on its own merit with no cajoling from the writer. Another phrase you should not use is

It is obvious . . .

If a remark is obvious, you shouldn't include it. If a remark isn't obvious, calling it "obvious" will only annoy your readers.

Many insecure scientists hide behind these arrogant phrases. These scientists think—subconsciously, perhaps—that these arrogant phrases will ward off challenges to their work. No such luck. If your research isn't strong, it shouldn't be published. The scientific world doesn't need weak research wasting space in its journals.

AVOIDING CLICHÉS

Clichés are figurative expressions that are too familiar. These expressions were once fresh in writing, but, through overuse, have taken on undesirable connotations.

Let's knock heads together *and figure out a solution.*

Most clichés are imprecise and unclear. Clichés are rampant in scientific correspondence and informal presentations.

We will touch base with you *on this report.*

What kind of time frame *are we talking about?*

When you come up to speed *on the computer, we will . . .*

The major thrust *of our penetrator program is to . . .*

This kind of writing is just plain silly. Unless you enjoy being laughed at, eliminate clichés from your writing.

REFERENCES

1. Thomas P. Johnson, "How Well Do You Inform?," *IEEE Trans. Prof. Com.*, Vol. 25, 1 (1982), p. 5.

CHAPTER 8

Being Concise

Vigorous writing is concise. A sentence should contain no unnecessary words, a paragraph no unnecessary sentences, for the same reason that a drawing should have no unnecessary lines and a machine no unnecessary parts. This requires not that the writer make all his sentences short, or that he avoid detail and treat his subjects only in outline, but that every word tell.[1]

William Strunk

Conciseness in language calls for eliminating redundancies and writing zeroes; it calls for reducing sentences to their simplest forms. Conciseness follows from pursuing two other language goals: clarity and forthrightness. When you make your writing forthright and clear, you also tighten it. Ridding sentences of pretentious diction such as "utilize" and "facilitate" leaves the more concise verbs "use" and "make." Ridding sentences of abstract nouns such as "factor" and "nature" cuts the fat prepositional phrases that accompany those nouns.

ELIMINATING REDUNDANCIES

Redundancies are needless repetitions of words. Redundancies either repeat the meaning of an earlier expression or else make points implicit in what has been stated.

The aluminum metal cathode became pitted during the experiment.

After "aluminum," the word "metal" is redundant.

The use of gaseous insulation is becoming increasingly more widespread.

The verb phrase "is becoming increasingly more widespread" is doubly redundant. A revision gives

Scientists are using gases more as insulators.

Below is a list of common redundancies in scientific writing. The words in parentheses should be deleted.

(already) existing	introduced (a new)
(alternative) choices	mix (together)
at (the) present (time)	never (before)
(basic) fundamentals	none (at all)
(currently) under way	now (at this time)
(completely) eliminate	period (of time)
(continue to) remain	(private) industry
(currently) being	(separate) entities
(empty) space	start (out)
(first) began	(still) persists
had done (previously)	whether (or not)

There are many more. Everyone writes redundancies in early drafts. You must catch redundancies in your editing. An effective way is to read through with the sole intention of cutting words—no additions allowed.

ELIMINATING WRITING ZEROES

Certain phrases have no meaning at all. They are zeroes in your writing—voids that offer no information to your readers.

It is interesting to note that *the K-alpha line of germanium . . .*

The phrase "it is interesting to note that" is a zero. If the fact about the K-alpha line of germanium isn't interesting, then you shouldn't include it. If the fact is important, more important than other results, then you should find a stronger way to signal its importance—perhaps with a one-sentence paragraph.

Sometimes, writing zeroes raise undesirable questions.

The requirements to be met for the measurement system include . . .

This sentence implies that there are requirements which will not be met. The phrase "to be met" is dangerously superfluous.

Some other common writing zeroes include

I might add	as a matter of fact
it should be pointed out	the fact that
it is significant	in a manner of
it is noteworthy	is used to
it is the primary intent of	the use of

Although these deletions may seem small, they are important. Eliminating zeroes doesn't just save a few minutes of reading time; it invigorates your writing. Concise writing is energetic writing.

REDUCING SENTENCES TO THEIR SIMPLEST FORM

Few scientists fail to reduce mathematical equations to their simplest form. Reducing equations to their simplest forms makes them easier to understand. The same is true with sentences.

Following observance of this occurrence, it was determined . . .

These eight words can be reduced to three:

We then determined . . .

Most scientific writing is full of fat. Fat slows the writing and tires the reader. Much fat in scientific writing arises from needless use of the passive voice:

The program being developed for fuels and chemicals applications can be divided *into two general categories. In the*

> *first category, applications* are being identified *which may, in the long term, be viable from an engineering and economic standpoint. In the second category, solutions* are being developed *to classes of engineering problems associated with coupling solar energy technology to a fuel or a chemical process.* (64 *words*)

Making the verbs active gives

> *The fuels and chemicals program* breaks down *into two categories. The first category* consists *of applications that may in the long term prove technically and economically feasible. The second category* consists *of solutions to engineering problems associated with coupling solar energy technology to a fuel or chemical process.* (48 *words*)

Not only does this revision save sixteen words; it also invigorates a dead paragraph.

Fat in scientific writing often arises from scientists avoiding the first person.

> *It was then concluded that a second complete solar mirror field corrosion survey should be conducted in July to determine if the tenfold annual corrosion rate projection was valid and to allow determination if subsequent corrective measures would be effective in retarding corrosion propagation.* (44 *words*)

A revision using the first person and our guidelines for precision gives

> *We conducted a second corrosion survey in July to see if the projected corrosion rate would continue and if stowing solar mirrors in the vertical position would slow the growth.* (30 *words*)

Another way to trim sentences is to cut fat phrases:

> In light of the fact that *costs for solar energy options have been cut in half over the past decade and future designs are even less expensive, solar energy could be an economical alternative to fossil fuels by the mid-1990s.*

The phrase "in light of the fact that" is wasteful; you should replace it with "because." Other common fat phrases in scientific writing and their reductions are given below:

at this point in time	now
at that point in time	then
has the ability to	can
has the potential to	can
in the event that	if
in the vicinity of	near
owing to the fact that	because
the question as to whether	whether
there is no doubt but that	no doubt

Many sentences acquire fat because scientists have converted verbs into nouns.

Our laboratory performed the research and development of . . .

A stronger, tighter sentence is

Our laboratory researched and developed . . .

Some other verbs whose energies are frequently lost in nouns are

perform a study	study
have a tendency	tend
make use of	use
make measurements of	measure
make a decision	decide
make a proposal	propose

Fat sentences make for slow reading, and most scientific papers—not some, not half, but most—are obese. It's difficult to trim sentences on early drafts when you're adding so much information. You must tighten your sentences on later drafts. You'll never eliminate all the fat in your writing, but you can reduce it. Most cuts are common sense. Try your hand at cutting fat from this preface written by the Department of Energy for one of its solar energy programs. The audience is varied: solar engineers, contract engineers, congressmen, the general public.

PREFACE

The research and development described in this document was conducted within the U.S. Department of Energy's

(DOE) Solar Thermal Technology Program. The goal of the Solar Thermal Technology Program is to advance the engineering and scientific understanding of solar thermal technology, and to establish the technology base from which private industry can develop solar thermal power production options for introduction into the competitive energy market.

The Solar Thermal Technology Program is directing efforts to advance and improve promising system concepts through the research and development of solar thermal materials, components, and subsystems, and the testing and performance evaluation of subsystems and systems. These efforts are carried out through the technical direction of DOE and its network of national laboratories who work with private industry. Together they have established a comprehensive, goal-directed program to improve performance and provide technically proven options for eventual incorporation into the Nation's energy supply.

To be successful in contributing to an adequate national energy supply at reasonable cost, solar thermal energy must eventually be economically competitive with a variety of other energy sources. Components and system-level performance targets have been developed as quantitative program goals. The performance targets are used in planning research and development activities, measuring progress, assessing alternative technology options, and making optimal component developments. These targets will be pursued vigorously to insure a successful program.[2]

This preface is full of redundancies, writing zeroes, and bloated sentences. Consider a revision with attention to our goals for language:

PREFACE

The research described in this report was conducted within the U.S. Department of Energy's Solar Thermal Technology Program. This program directs efforts to incorporate technically proven and economically competitive solar thermal options into our nation's energy supply. These efforts are carried out through a network of national laboratories that work with industry.

Although this revision is less than half the length of the original, it has more than twice the power. Fat writing is lethargic writing. Concise writing moves.

REFERENCES

1. William Strunk and E. B. White, *Elements of Style,* (New York: Macmillan Publishing Co., 1979), p. 23.
2. U. S. Department of Energy, *Solar Thermal Technology Annual Evaluation Report,* (Golden, CO; Solar Energy Research Institute, 1985), p. ii.

CHAPTER 9

Being Fluid

It don't mean a thing if it ain't got that swing.

Louis Armstrong

Many scientists mistakenly believe that scientific writing must be dull. Unfortunately, their writing reflects that misconception. It is dull, needlessly dull. Many scientists undercut the purpose of scientific writing—to inform—with sentences and paragraphs that drag, with discontinuities in language that trip readers. Scientists study the most fascinating subjects in the world: space exploration, animal and plant kingdoms, the inner workings of atoms and nuclei. Why then do so many scientists resign themselves to prose styles that are sluggish, that are without life?

No one expects you to write with the grace of John Cheever or Joyce Carol Oates, but you do have to inform. If you're going to inform, you can't bore readers with monotonous language—language without variety, language full of discontinuities. Your language must be precise and clear. It must be anchored in the familiar. It must be forthright and concise, and it must be fluid. Fluid language is the lubricant that makes your writing inform. Although the two constraints of scientific writing somewhat limit stylistic variation, scientific language can still be fluid. It can be energetic, sometimes even exciting.

VARYING SENTENCES

Many scientific papers read slowly because the sentences have no variety. The sentences begin the same way. They have the same length. They have the same arrangement of nouns, phrases, and verbs. In any type of writing, whether it be a poem or feasibility report, there are rhythms. Rhythms determine the energy of the writing. Writing that uses the same rhythms over and over is dull reading. Imagine a piano piece with only two or three different notes. Not very exciting, huh? But that is the way many scientists write—two or three sentence patterns repeated again and again. The result is stagnation.

> *Mount St. Helens erupted on May 18, 1980. A cloud of hot rock and gas surged northward from its collapsing slope. The cloud devastated more than 500 square kilometers of forests and lakes. The effects of Mount St. Helens were well documented with geophysical instruments. The origin of the eruption is not well understood. Volcanic explosions are driven by rapid expansion of steam. Some scientists believe the steam comes from groundwater heated by magma. Other scientists believe the steam comes from water originally dissolved in the magma. We need to understand the source of steam in volcanic eruptions. We need to determine how much water the magma contains.*

The subject matter is interesting, but the prose is tiresome, and the rhythms monotonous.

How do you vary sentence rhythm? One way is to *vary the way sentences begin*. Some common ways to begin sentences include

Subject-Verb:

Mount St. Helens erupted *on May 18, 1980.*

Verb Phrase:

Its slope collapsing, *the mountain emitted a cloud of hot rock and gas.*

Prepositional Phrase:

In a matter of minutes, *the cloud devastated more than 500 square kilometers of forests and lakes.*

Introductory Clause:

Although the effects of the eruption were well documented, *its origin is not well understood.*

Adverb:

Volcanic explosions are driven by a rapid expansion of steam. Recently, *debate has arisen over the source for the steam.*

Verb Introducing Question:

Is *it groundwater heated by magma or water originally dissolved in the magma itself?*

Infinitive Phrase:

To understand the source of steam in volcanic eruptions, *we need to determine how much water the magma contains.*

Using these sentence patterns, we revise our paragraph to

Mount St. Helens erupted on May 18, 1980. Its slope collapsing, the mountain emitted a cloud of hot rock and gas. In a matter of minutes, the cloud devastated more than 500 square kilometers of forests and lakes. Although the effects of the eruption were well documented, its origin is not well understood. Volcanic explosions are driven by a rapid expansion of steam. Recently, debate has arisen over the source for the steam. Is it groundwater heated by magma, or water originally dissolved in the magma itself? To understand the source of steam in volcanic eruptions, we need to determine how much water the magma contains.[1]

Now don't think you must use these sentence openers in any particular pattern. Strong writing contains no magic formulas for sentence openers (or any other aspect of style). The way you vary sentence openers determines your individual style. Just remember that failure to vary sentence openers will stagnate your prose and exhaust your readers.

Another way to vary sentences is to *vary sentence length.* In most scientific papers, sentences are too long—they average over twenty-five words. Compare that average length to the average length in *Newsweek*, which is only about seventeen words. With sentence variety, the average length of your sentences is not as important as how your sentence length vary. What stagnates many scientific papers is that too many sentences have the same length, often between twenty and twenty-eight words.

> *On the morning of May 18, a strong earthquake shook the volcano causing its cracked and steepened north side to slide away. (22) Photographs taken during these early seconds, together with other information, showed thai the blast originated 500 meters beneath a bulge on the north face. (24) Photographs and time of destruction of a seismic station established the velocity of the blast at about 175 meters per second. (21) Significantly, the volume of new magmatic material ejected in the blast (about 0.1 km³) appears to equal the volume of the bulge. (22)*

This paragraph is readable, but the sentence length distribution is narrow (21–24 words), making the writing monotonous.

What distribution should you use? A Gaussian distribution? A Bose distribution? The answer is no one knows. Here are some general guidelines for fluid writing.

1. Change your sentence length every two or three sentences.
2. Keep your average length in the teens.
3. Occasionally use a very short or very long sentence.

Don't take these guidelines out of context. In your writing, you must juggle six other language goals besides fluidity.

How short can you make your sentences? You can make your sentences as short as you want. An occasional three- or four-word sentence can accent an important result, particularly if the short sentence follows a rather long sentence. Be careful with short sentences though; too many short sentences will make your writing choppy. How long can a sentence be? As long as you want, provided that you maintain clarity.

On the morning of May 18, a strong earthquake shook the volcano causing Mount St. Helens' cracked and steepened north side to slide away. (24) Photographs showed that the blast began beneath a bulge on the north face. (13) These photographs when coupled with other information established the depth of the blast's origin to be 500 meters and the velocity of the blast to be about 175 meters per second. (31) The measured volume of the ejected magmatic material was about 0.1 km^3. (11) This volume appears to equal the volume of the bulge. (10)

A third way to vary sentences is to *vary the sentence structure*. The three basic sentence structures are *simple sentences, complex sentences*, and *compound sentences*. Simple sentences are sentences with one subject and one verb.

Moreover, lava flows *(subject) from nonexplosive eruptions ordinarily* contain *(verb) only 0.2% water.*

Complex sentences are simple sentences joined with dependent clauses.

Although the amount of devastation caused by the May 18 blast was a surprise, *(dependent clause) the eruption itself had been expected for weeks.*

Compound Sentences are two or more simple sentences joined by a conjunction.

Precursor activity to the eruption began on March 20, 1980, and *(conjunction) many times during the next two months the mountain shook for minutes.*

There are other kinds of structures, such as compound-complex, but don't worry about them. It's not really important that you know all the kinds of sentence structures or even the ones mentioned here. What is important is that your sentence structures don't stagnate. Again, there are no formulas for varying sentence structure, no ratios for simple to compound sentences. The advice is this: If your sentence structures start to look the same, vary them.

Conscious variation of sentence openers, sentence lengths, and sentence structures is difficult during early drafts. During early drafts, you must concentrate on getting sentences on the paper, not on how many simple sentences in a row you have. Don't worry about sentence variety until the polishing drafts. Also, don't forget that sentence variety is secondary to sentence clarity. If a choice arises between fluidity and clarity, choose clarity.

VARYING PARAGRAPHS

Before we talk about varying paragraphs, let's establish what a paragraph is. A *paragraph* is a group of sentences that conveys a group of connected ideas. The particular way you group ideas is individual; it is part of your individual style. Just remember that the same philosophy about varying sentences applies to varying paragraphs. If your paragraphs don't vary in size and structure, your writing will drag.

Varying your paragraph size. There's no rule for how long a paragraph should be. In a complex scientific paper, readers will typically tire after about fifteen lines. There's also no rule about how short a paragraph can be—one, two, or three lines if you desire. A short paragraph is an excellent way to accent an important result, particularly if the short paragraph follows a long one. Be careful; too many short paragraphs, just like too many short sentences, will make your writing choppy.

Varying your paragraph structure. The number, length, and type of sentences in a paragraph determines its structure. As with sentence structure, the important thing to remember is that repetitive paragraph structures slow the reading. While editing a paper, if you find a particular section tiresome, look at the paragraph structures. Maybe you began all your paragraphs with simple sentences introduced by prepositional phrases. Maybe all your paragraphs have five sentences. If precision and clarity allow for it, switch things around.

Many scientists cling to the misconception that every paragraph should begin with a thesis sentence. Well, that's silly. Few good writers, if any, could give you a definition of a thesis sentence, and even fewer writers consciously begin their paragraphs with one. What is important is that the opening sentence of a paragraph connects the ideas of that paragraph to the ideas of the preceding paragraphs. Opening sentences to paragraphs must provide transition—transition from one group of ideas to another. Whether the transition is made with a *thesis sentence* (whatever that is), or some other kind of sentence, is unimportant.

One English teacher I had demanded that you use five sentences—no more, no less—in each paragraph, and that each of those five sentences be structured a particular way. The first sentence had to be a thesis sentence that listed three ideas, the middle three sentences then discussed each of those ideas, and the last sentence was always a wrap-up sentence for the paragraph. Furthermore, this teacher limited you to five paragraphs for each paper, no matter what the subject. That kind of strategy might work for two pages on the symbolism of Doctor T. J. Eckleburg's eyeglasses in *The Great Gatsby*, but would exasperate a reader of a twenty-page paper on the "Monte Carlo Simulation of Rocket Plume Enhancement Regions."

ELIMINATING DISCONTINUITIES

Poor transitions between sentences and paragraphs cause discontinuities in language. Some of these discontinuities arise from too much distance between connected words and thoughts.

> *The Cascade Range, with its prominent chain of towering cones, is not the only threatening volcanic region in the western United States. Many people who live in the eastern Sierra Nevada community of Mammoth Lakes, California, may have been unaware until recently that their scattered hills and ridges have a remarkably recent volcanic origin as well.*

Not until the last four words of the second sentence is there a tie between these two sentences. The geologist keeps her readers in limbo for over thirty words before making a connection. That's too long. Use connective words and phrases early in sentences to make strong transitions between sentences.

The Cascade Range, with its prominent chain of towering cones, is not the only threatening volcanic region in the western United States. The Mammoth Lakes area of the Sierra Nevada also has a remarkably recent volcanic origin.

The word "also" ties the second sentence in with the first. Words such as "also" are connectives. Some other connectives include

and	moreover
but	next
finally	therefore
however	then

Other causes for poor transitions in scientific writing are gaps in logic, places where key details have been omitted. The writer doesn't trip over the gaps because the writer knows all the key details, but the reader trips. In the Cascade Range paragraph, the geologist assumed her readers knew that the Sierra Nevada was a separate mountain range from the Cascade Range. This assumption depended on the audience. If the audience was diverse, the paragraph would better read

The Cascade Range, with its prominent chain of towering cones, is not the only threatening volcanic region in the western United States. The Mammoth Lakes area of the Sierra Nevada Range also has a remarkably recent volcanic origin.

or, better still,

The Cascade Range, with its prominent chain of towering cones, is not the only threatening volcanic region in the western United States. Further south, the Mammoth Lakes area of the Sierra Nevada Range also has a remarkably recent volcanic origin.

The word "Range" after "Sierra Nevada" and the phrase "Further south" strengthen the connection between the two sentences.

In derivations of equations, discontinuities often arise from weak transitions from between steps. Although you must often compress derivations, you shouldn't take needless jumps.

Our equation for intensity I now is

$$I = \int_{-1}^{+1} \frac{dx}{\sqrt{1 - x^2}(1 + x^2)} \, .$$

Using elementary techniques, we then arrive at

$$I = \frac{\pi}{\sqrt{2}} \, .$$

This physicist not only lost continuity in the derivation but also infuriated many readers who didn't *immediately* recognize how to integrate the equation. For almost the same numbers of words, the physicist could have written

Our equation for intensity I now is

$$I = \int_{-1}^{+1} \frac{dx}{\sqrt{1 - x^2}(1 + x^2)} \, .$$

Using a contour integral, we then arrive at

$$I = \frac{\pi}{\sqrt{2}} \, .$$

Stating how the equation was solved cost the physicist nothing, but possibly saved his readers much work.

Discontinuities can also arise from nonparallel wording.

The system must be reliable, it must meet power requirements, and have high efficiency.

The last clause "have high efficiency" is not parallel to the first two clauses because it has no noun or pronoun. Make your lists parallel.

The system must be reliable, it must meet power requirements, and it must have high efficiency.

Discontinuities can also arise from abbreviations.

Long Valley near Mammoth Lakes, CA, is a caldera that was formed 700,000 yrs ago, i.e., the roof of a huge magma chamber collapsed.

Eliminate abbreviations that don't make the writing more concise. In scientific writing, concise writing is writing that takes your readers the shortest time to understand.

> *Long Valley, near Mammoth Lakes, California, is a caldera that was formed 700,000 years ago when the roof of a huge magma chamber collapsed.*

Use

versus	instead of	vs.
meaning that	instead of	i.e.
for example	instead of	e.g.

Your writing will read much more smoothly.

REFERENCES

1. John C. Eichelberger, "Modeling Mount St. Helens Volcanic Eruption," *Sandia Technology*, 7, no. 2 (June 1983), p. 3.

CHAPTER 10

Being Imagistic

Go in fear of abstractions.

Ezra Pound

Pound's advice to poets has merit in scientific writing. People think and remember images, not abstractions. Therefore, you need to anchor your writing with images. Otherwise, your readers will forget your research. Let's assume you helped design a "particle beam fusion accelerator," an accelerator that focuses particle beams on deuterium-tritium pellets in an attempt to produce nuclear fusion. Now, when you write the phrase "particle beam fusion accelerator," you see a specific image. You see a specific image because you helped design the accelerator. Maybe you see something such as Figure 10-1, a cutaway view of the whole experiment: the huge capacitors submerged in oil, the circular diode that produces the particle beams, and the inner chamber where the beams converge on the pellets. Your readers, however, don't hold this image. Your readers see only the phrase "particle fusion beam accelerator." Therefore, in your writing you must somehow supply an image for this phrase.

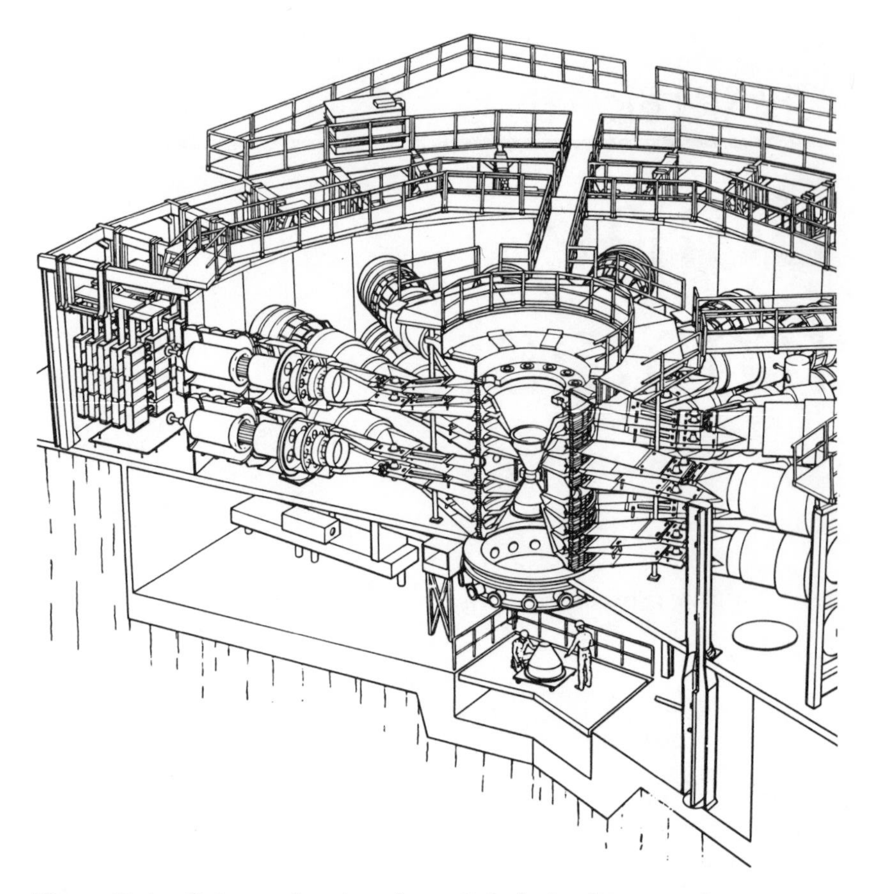

Figure 10-1. Cutaway drawing of a particle fusion beam accelerator. This accelerator focuses beams of lithium ions onto deuterium-tritium pellets in an attempt to produce nuclear fusion.[1]

One way to provide images is through illustration (which is discussed later). However, illustration is not enough; your writing must supplement illustration with imagistic language. Scientific language, at its strongest, makes readers see the research. Einstein was particularly adept at using images. Although many ideas of relativity were abstract, Einstein always anchored his arguments with common images: ships passing, birds flying, stones dropping from trains.

Using imagistic language demands imagination. More

than that, it demands hard work. You must place yourself in the eyes of the reader, then walk through your research and see things as though for the first time. To make your language imagistic, you must present details so readers can see what you've seen, think what you've thought.

USING CONCRETE DESCRIPTION

Concrete description is specific description; description that uses the five senses: sight, sound, taste, touch, smell.

> *Quicksand is a thick bed of fine sand, as at the mouth of a river or along a seacoast, that consists of smooth rounded grains. The bed is saturated with water flowing upwards through voids; thus, the bed is a soft shifting mass that yields easily to pressure and tends to suck down and readily swallow heavy objects on its surface.*

Notice the textures described in this definition: "thick bed," "fine sand," "smooth rounded grains," "saturated with water," "soft shifting mass." This definition doesn't just tell readers what quicksand is; it makes readers feel quicksand.

SELECTING IMAGISTIC WORDS

Select words that give pictures.

> *In Bohr's theory of the hydrogen atom, electrons* orbit *nuclei.*

The verb "orbit" suggests the physical image of an electron moving around a nucleus the same way the earth moves around the sun. This kind of writing, when selectively used, is effective. Another example:

> *The unique feature of their design is a continuous manifold, which follows a unidirectional—as opposed to serpentine—flow for the working fluid.*

The word "serpentine" suggests the image of a coiled serpent. "Serpentine" supplies readers with an image; the word "unidirectional" does not.

Be careful with imagistic words. You can easily mix images that are not compatible.

> *In the breakdown of a gas, "seed" electrons, which are continuously "emitted" from the avalanche, multiply at various distances from the parent avalanche, thus rapidly expanding the avalanche space charge toward the anode.*

The imagination behind this sentence is good; however, the number and variety of images work against each other. The writer infuses "seed" images with "parent" images against a backdrop of "avalanche" images. This combination is too confusing.

USING IMAGISTIC ANALOGIES

Analogies compare obtuse thoughts to familiar ones. Analogies allow readers to see the writer's thoughts. Einstein used them generously in his writing:

> *I stand at the window of a railway carriage which is travelling uniformly, and drop a stone on the embankment, without throwing it. Then, disregarding the influence of air resistance, I see the stone descend in a straight line. A pedestrian who observes the misdeed from the footpath notices that the stone falls to earth in a parabolic curve. I now ask: Do the "positions" traversed by the stone lie "in reality" on a straight line or on a parabola?*[2]

Einstein's analogy is so much more alive than the abstract question: Where do positions of an object lie in reality? Unfortunately, many scientific papers are void of analogies. Analogies are tools to help readers. Analogies demand imagination and creativity. The best scientists use them. Analogies not only provide insights into how you think; they also bring life into your writing.

REFERENCES

1. Pace VanDevender, "Ion-Beam Focusing: A Step Toward Fusion," *Sandia Technology*, 9, no. 4 (December 1985), pp. 2–13.

2. Albert Einstein, *Theory of Relativity*, (London: London, Methuen & Co., LTD, 1950).

PART THREE
Illustration

CHAPTER 11

The Meshing of Words with Pictures

The most beautiful thing we can experience is the mysterious. It is the source of all true art and science.

Albert Einstein

Illustration is not just the presence of figures and tables. Illustration is the meshing of figures and tables with language. Just pasting pictures into your paper is not strong scientific writing. In fact, a slapped-in figure or table may confuse your readers more than inform them.

To communicate scientific research, you need effective illustrations. Illustrations can clarify images too complex to be conveyed by language.

> *The new solar mirror design consists of a circular support ring with a hollow rectangular cross-section. The ring has identical prestressed front and rear membranes that are attached with full perimeter welds. A polymer film is bonded to the front membrane, and internal pressure between the membrane controls the focal length.*

In this description, the engineer demands too much from his readers. His readers have to imagine the support that holds the ring as well as the ring's hollow rectangular cross-section, as well as the relationship between the front and rear membranes that produce a constant focal length.

There is nothing wrong with the language in the above description; it's just that the engineer needs an illustration to convey the image to his readers.

> *The new solar mirror design, shown in Figure 11-1, consists of a circular support ring with a hollow rectangular cross-section. The ring has identical prestressed front and rear membranes that are attached with full perimeter welds. A polymer film is bonded to the front membrane, and internal pressure between the membranes controls the focal length.*

In this second description, the figure clarifies the image. Although the second description is longer, the writer makes sure the reader sees the image. Note that just placing in the

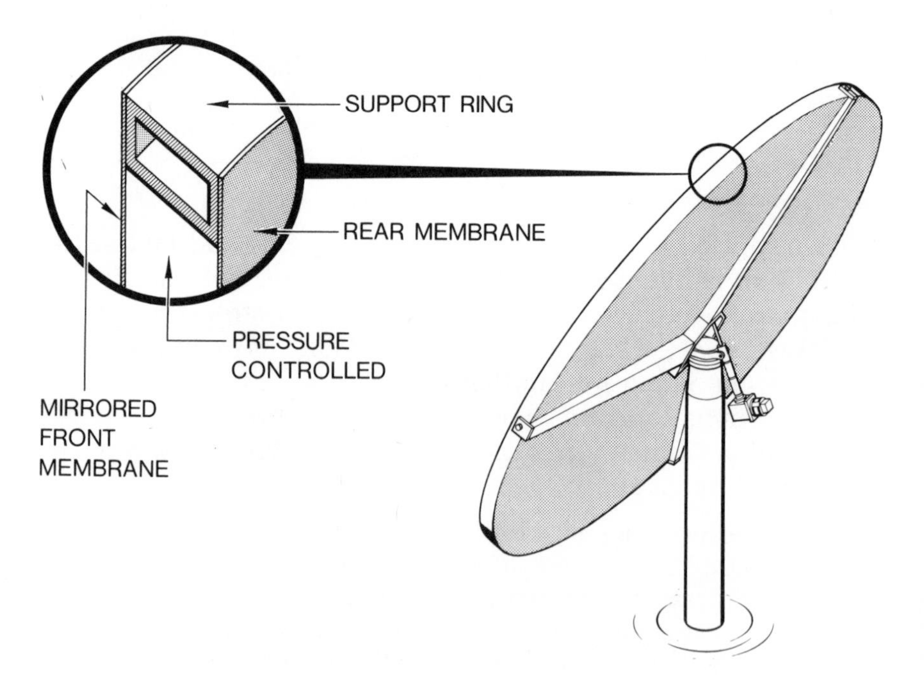

Figure 11-1. New solar mirror design.[1]

figure without the accompanying text would not have been effective.

> *The new solar mirror design is shown in Figure 11-1. Internal pressure between the membranes controls the focal length.*

This description does not explain the different parts of the mirror design. There is not enough transition between words and picture.

Besides clarifying images, illustrations also give readers rest stops. Readers of scientific papers need rest stops so complex details can soak in. Moreover, handsome illustrations such as Figure 11-2 can reduce the complex appearance of scientific papers. A scientific paper packed with text and

Figure 11-2. Solar furnace at Odeillo, France. Mirrors concentrate solar energy onto the furnace to attain temperatures up to 4000°C.[2]

equations intimidates readers. Illustrations can make scientific papers more palatable. Don't assume, however, that you should fill your papers with graphics the same way a comic book is filled with pictures. Too many illustrations will reduce the efficiency of informing. As with language, the way you illustrate depends on your audience and your research.

REFERENCES

1. C. L. Mavis, "Stressed Membrane Heliostat," *Solar Central Receiver Program Bimonthly Report (October-November)*, (Livermore, CA: Sandia National Laboratories, 1984), p. 5.
2. F. Trombe, et. al., "Thousand kW Solar Furnace—Built by the National Center of Scientific Research in Odeillo (France)," *Solar Energy*, 15 (1973), pp. 57–61.

CHAPTER 12

Types of Illustrations

There is no Heaven but clarity, no Hell except confusion.

Jan Struther

There are two types of illustrations: tables and figures. Tables are arrangements of numbers and descriptions in rows and columns. Figures are everything else: photographs, drawings, diagrams, graphs.

TABLES

Tables have two important uses. First, tables can present numerical data with a high degree of accuracy. With a table such as Table 12-1, you can present data with as many significant digits as you desire.

Besides presenting numerical data with a high degree of accuracy, tables can also present parallel descriptions. Don't think that tables are meant only for numbers. Many scientists neglect to take advantage of the power that tables have to concisely present parallel descriptions that would otherwise have to be listed in tedious sentences.

Solar Power Plant Options

We have studied four solar central receiver designs for a 50 MW_e power plant. The design characteristics of the four options are given in Table 12-2.

All four options have a 50 MW_e plant rating and use a flat plate receiver with north field mirror configuration. Options 1 and 2 are stand-alone plants, while Options 3 and 4 are hybrid plants (plants that use fossil fuels to back up the solar technology). Options 1 and 3 use molten salt as the heat transfer fluid in the receiver, while Options 2 and 4 use liquid sodium. All four options use molten salt as the heat transfer fluid in the thermal storage tank.[1]

Without Table 12-2, this discussion would be much longer and more complex. Moreover, Table 12-2 allows readers to compare quickly the designs of each option.

Table 12-1
Blood-Glucose Levels for a Normal Individual and a Diabetic

Time (hour)	Normal (mg/dl)*	Diabetic (mg/dl)
12:00 a.m.	100.3	175.8
2:00	93.6	165.7
4:00	88.2	159.4
6:00	100.5	72.1
8:00	138.6	271.0
10:00	102.4	224.6
12:00 p.m.	93.8	161.8
2:00	132.3	242.7
4:00	103.8	219.4
6:00	93.6	152.6
8:00	127.8	227.1
10:00	109.2	221.3

*(milligrams/decaliter)

Table 12-2
Design Characteristics of Solar Power Plant Options

	Option 1	Option 2	Option 3	Option 4
Plant Type	Stand-Alone	Stand-Alone	Hybrid	Hybrid
Plant Rating (MW_e)	50	50	50	50
Field Configuration	North	North	North	North
Receiver Shape	Flat Plate	Flat Plate	Flat Plate	Flat Plate
Receiver Fluid	Salt	Sodium	Salt	Sodium
Receiver Output Power (MW_t)	225	210	225	210
Receiver Temperature (°C)	574	574	574	574
Intermediate Heat Exchanger	None	Sodium-Salt	None	Sodium-Salt
Thermal Storage Fluid	Salt	Salt	Salt	Salt
Storage Capacity (MW_t-hour)	450	450	200	200

FIGURES

Figures include photographs, drawings, diagrams, and graphs.

Photographs

Photographs give readers realistic depictions of images and events. The major advantage of photographs is realism; photographs present readers with exact details. This advantage is also a disadvantage. A photograph not only shows a subject's true textures and colors, but also its true shadows and scratches. These extraneous details can easily confuse readers. Consider Figure 12-1, a photograph of the Solar One Power Plant. Although the photograph is impressive, it also raises an important question: What are the streaks of light on

either side of the tower? If you use this figure, you should explain those streaks.

> *The bright streaks on either side of the receiver are* standby points, *safety targets for the mirror beams to prevent them from converging in the airspace above the plant and blinding pilots. We first focus mirrors on these standby points whenever we start up or shut down the plant.*

If you do not have room to explain the streaks, then you should not use this photograph.

Drawings

Drawings include line sketches and artists' renditions. The major advantage of drawings is that you can control the amount of precision. Unlike photographs, if there's a detail that does not enhance the discussion, you can easily delete it. Figure 12-2, for example, is a steam generator for a solar power plant. With photography, you could not isolate this generator from other parts of the plant. The plant's tower, piping, and storage tanks would get in the way. Besides allowing you to delete extraneous details, drawings also permit unique perspectives such as cutaway, blow-up, and exploded. Figure 12-3 is a cutaway of a nuclear fusion experiment. The inner detail of this experiment would be impossible to attain with photography. Finally, drawings allow you futuristic perspectives. Figure 12-4 is an artist's rendition of a satellite firing a probe into the comet Kopff. Kopff will pass the earth in 1996.

Diagrams

Diagrams are drawings that communicate through symbols and do not try to depict an object's physical characteristics. Figure 12-5, for example, shows the energy flow through the nuclear fusion experiment of Figure 12-3. A common type of diagram is a schematic. Figure 12-6 is a schematic

---

Figure 12-1. Solar One Power Plant located near Barstow, California. Mirrors focus solar energy onto a central receiver where water is converted to steam. The steam powers a turbine to produce 10 megawatts to a utility grid.[2]

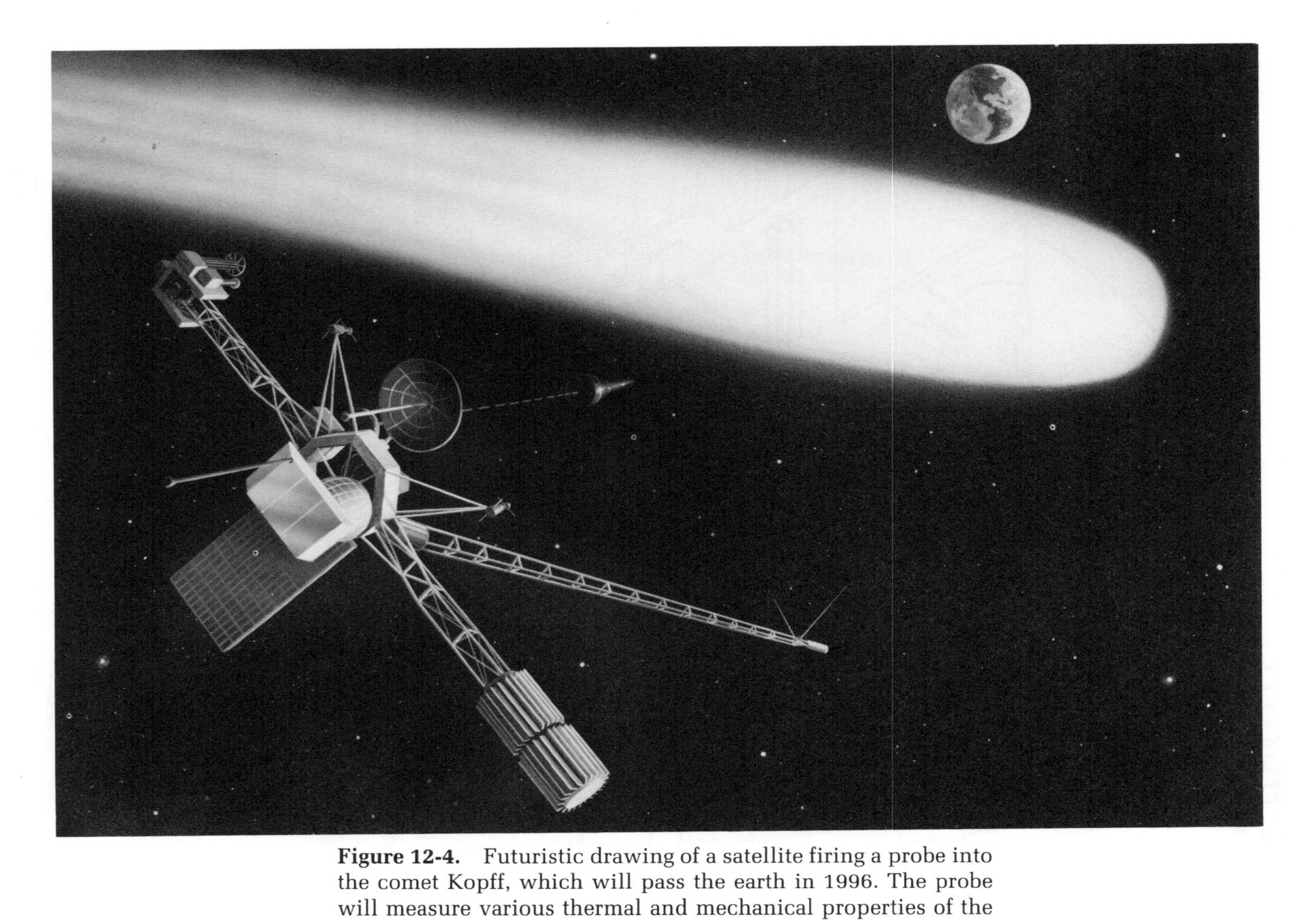

Figure 12-4. Futuristic drawing of a satellite firing a probe into the comet Kopff, which will pass the earth in 1996. The probe will measure various thermal and mechanical properties of the comet.[5]

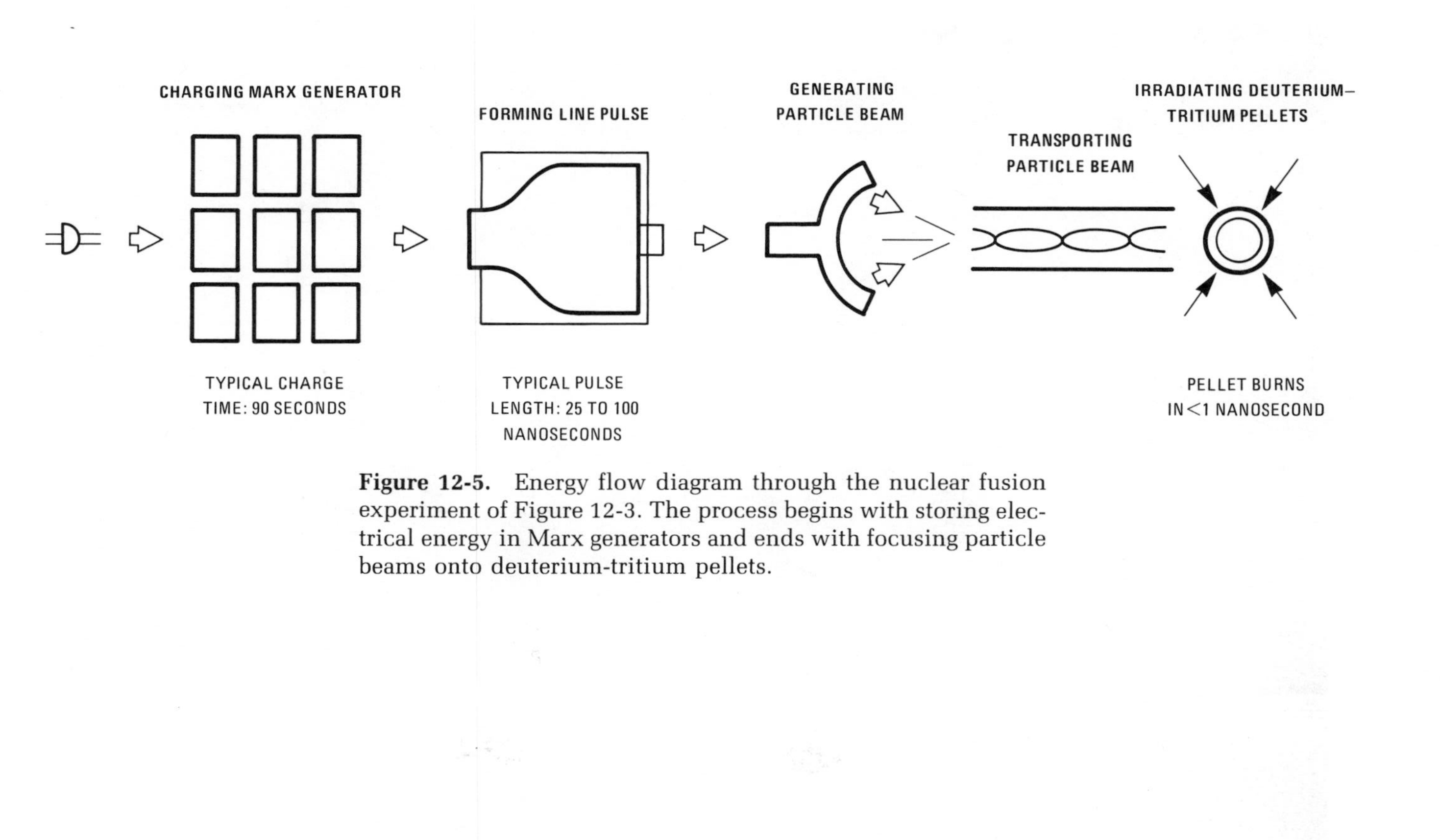

Figure 12-5. Energy flow diagram through the nuclear fusion experiment of Figure 12-3. The process begins with storing electrical energy in Marx generators and ends with focusing particle beams onto deuterium-tritium pellets.

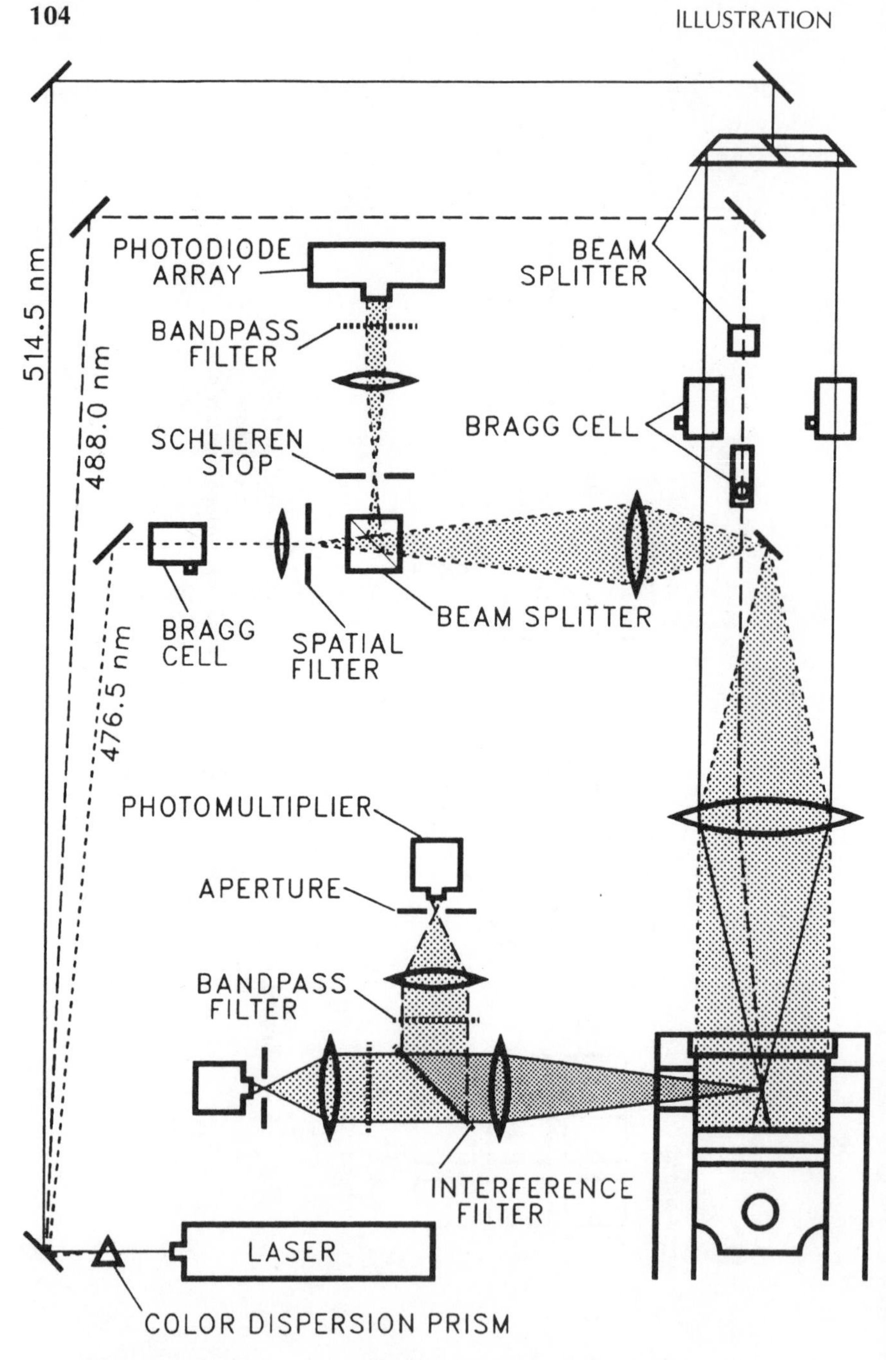

Figure 12-6. Schematic of a laser system used to measure the flame speed, flame position, and cylinder pressure in an internal combustion engine.[6]

of a laser system used to measure flame speed, flame position, and cylinder pressure in an internal combustion engine. Schematics show how different parts of a system relate to one another. When using a schematic, you must make sure your readers know what the symbols in your schematic mean. Otherwise, your schematic won't inform.

Graphs

Graphs are drawings that show general relationships in data. There are three common types of graphs: line, bar, and circle. Line graphs are the most common type of graph in scientific writing. Line graphs are effective at showing general trends. If you desire a high degree of accuracy, you should use a table. Figure 12-7 presents the blood-glucose levels of a nor-

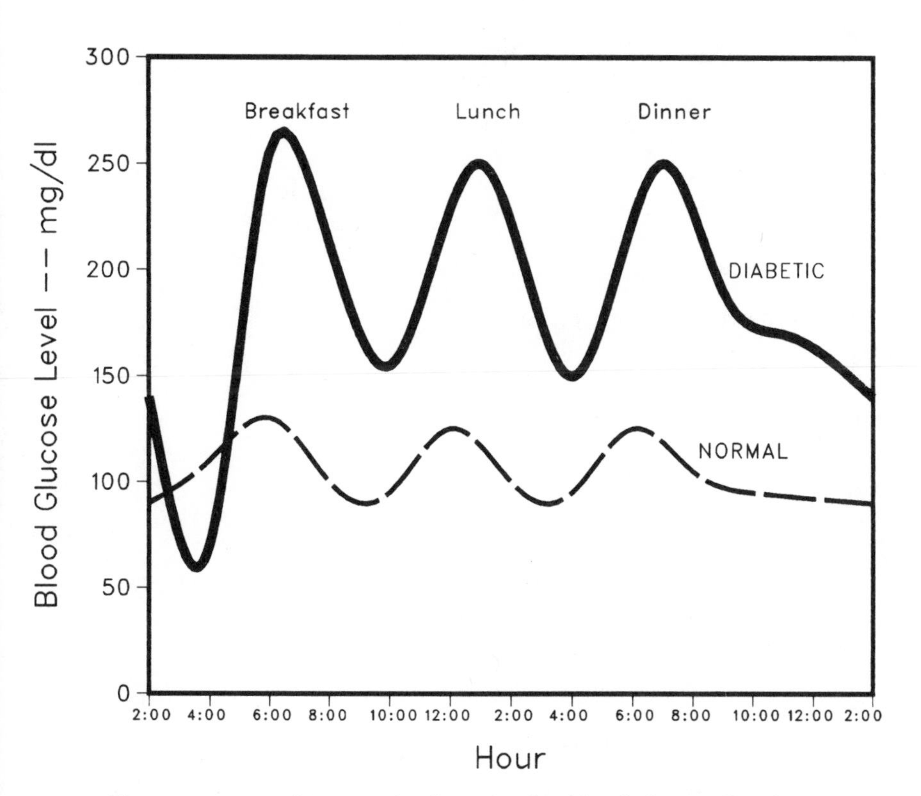

Figure 12-7. A line graph showing the blood-glucose levels over a twenty-hour period for a normal individual and for a diabetic. Levels are in milligrams/decaliter (mg/dl).[7]

mal individual and a diabetic plotted over a twenty-four hour period. Notice that Figure 12-7 presents the same information as found in Table 12-1. Whereas you attain more accuracy in tables, you show general trends more clearly in graphs.

Circle graphs allow you to compare parts of a whole. Figure 12-8, for example, breaks down the earth's surface area covered by ocean. While circle graphs compare parts of the same whole, bar graphs compare sizes of different wholes. Figure 12-9 shows the relative sizes of the earth's seven continents.

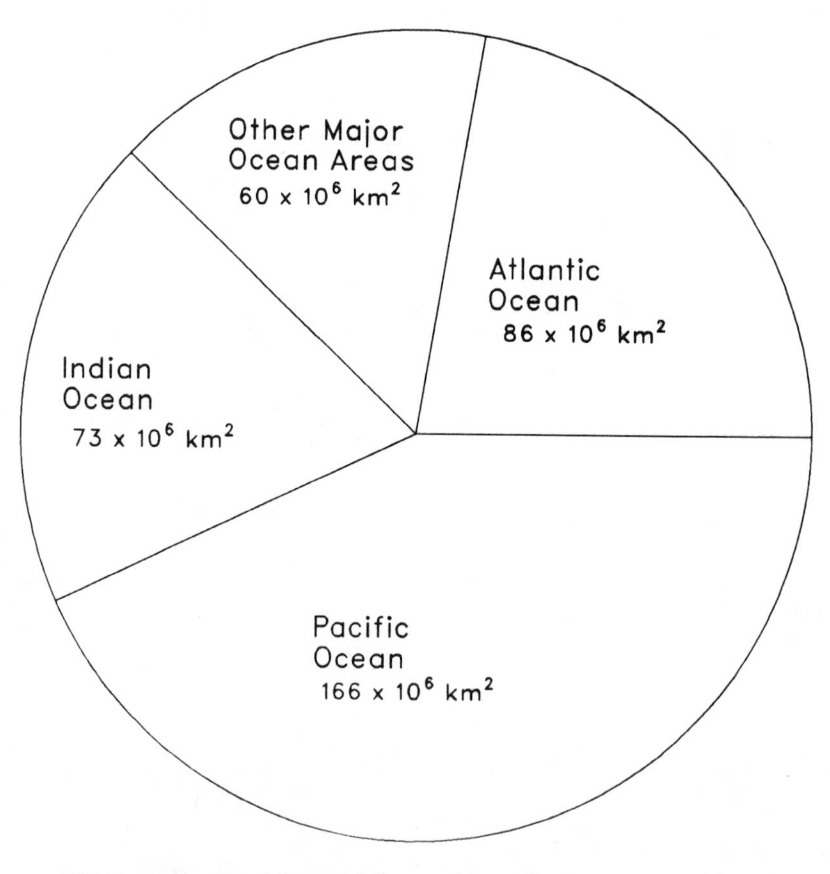

Figure 12-8. Breakdown of the earth's surface area covered by ocean (total surface area equals 375×10^6 km^2).

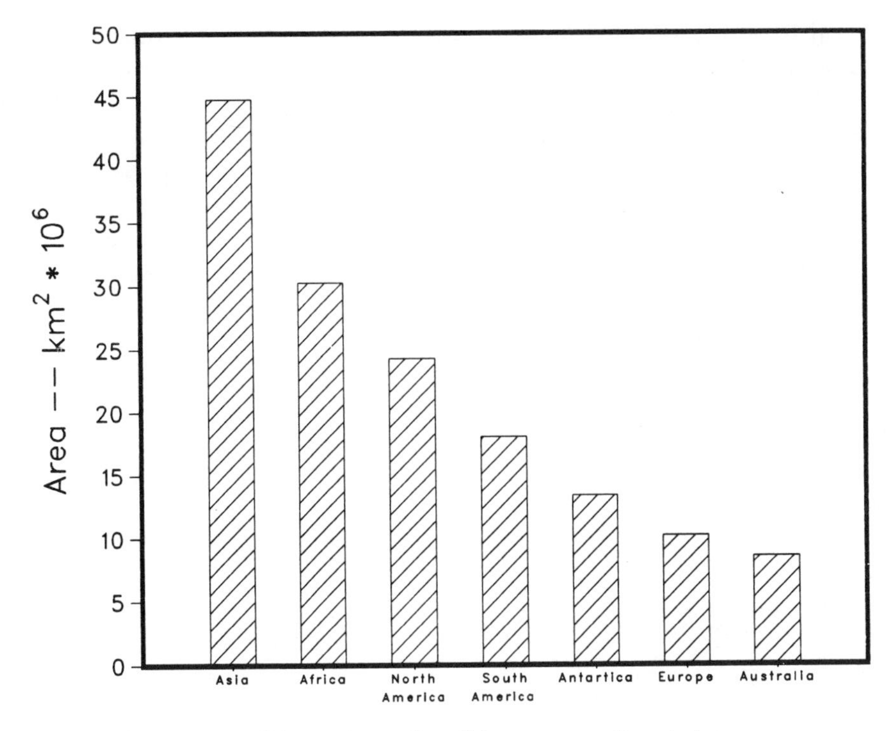

Figure 12-9. Dimensions of earth's seven continents in square kilometers.

REFERENCES

1. P. K. Falcone, *A Handbook for Solar Central Receiver Design*, SAND86-8006, (Livermore, CA: Sandia National Laboratories, 1986).
2. L. G. Radosevich, *Final Report on the Experimental Test and Evaulation Phase of the 10 MWe Solar Thermal Central Receiver Pilot Plant*, SAND85-8015, (Livermore, CA: Sandia National Laboratories, 1986).
3. Jack Young, "Molten Salt Electric Experiment Steam Generator Subsystem," SL-20488a, (Livermore, CA: Sandia National Laboratories, 1985).
4. Pace VanDevender, "Ion-Beam Focusing: A Step Toward Fusion," *Sandia Technology*, 9, no. 4 (December 1985), pp. 2–13.

5. Jack Young, "Satellite Firing Probe into Comet Kopff," SL-21518 (Livermore, CA: Sandia National Laboratories, 1985).
6. P. Witze, "Experimental Facility for Multiparameter Conditionally Sampled Laser Velocimetry," Presented at *Proceedings of the International Symposium on Diagnostics and Modeling of Combustion in Reciprocating Engines*, (Tokyo, Japan: 1985).
7. Gary Carlson, "Implantable Insulin Delivery System," *Sandia Technology*, 6, no. 2 (June 1982), pp. 12–21.

CHAPTER 13

Goals of Illustrations

Everything should be as simple as it can be, yet no simpler.

Albert Einstein

The constraints on illustrations are the same as on language:

1. You must inform your audience as efficiently as possible.
2. You must stay honest.

To meet these constraints for illustrations, you should pursue the same goals as you did for language. First, you must make your illustrations *precise*. Precision does not mean having the most accurate illustrations. Precision means having illustrations that best reflect the degree of accuracy in your language. An equally important goal of illustrations is being *clear*. Too often in scientific writing, illustrations confuse rather than inform. Clarity demands not only that your illustrations be understood, but also that they not be misunderstood.

Besides striving for clarity and precision, you should anchor your illustrations in the *familiar*. If your research produces illustrations with unusual images, then you should explain the origin of those images. Illustrations should also be *forthright;* they should be straightforward and honest—no sleight-of-hand manipulations of data.

Because of the avalanche of scientific research published every year, you must make your illustrations *concise*. Conciseness does not necessarily mean the smallest-spaced illustration—rather, the illustration that informs your reader in the shortest time. Your illustrations should also be *fluid*. You need to make smooth transitions between words and pictures. Finally, just as your langauge should be *imagistic*, so should your illustrations. Don't think that having imagery as a goal for illustrations is a redundancy. The mere presence of illustrations in your paper is no guarantee that your writing is imagistic. Many types of illustrations such as tables and diagrams do not contain images, and many photographs and drawings in scientific papers do not present the most important images of the research.

BEING PRECISE

Einstein said to keep your information as simple as possible, yet no simpler. This advice particularly applies to choices of precision in illustrations. In your illustrations, don't include details that aren't self-explanatory or explained in your text. The precision of your illustrations should reflect the precision of your language. A common mistake in scientific writing is to present a figure that is much more complex than the prose.

> *The thermal storage system stores heat in a huge, steel-walled insulated tank. Steam from the solar receiver passes through heat exchangers to heat the thermal oil, which is pumped into the tank. The tank then provides energy to run a steam generator to produce electricity. A schematic of this system is shown in Figure 13-1.*

Figure 13-1 is just too complex for the prose given. Perhaps you could include this schematic in an appendix for readers already familiar with the system, but in the general discussion, you should use a figure such as Figure 13-2.

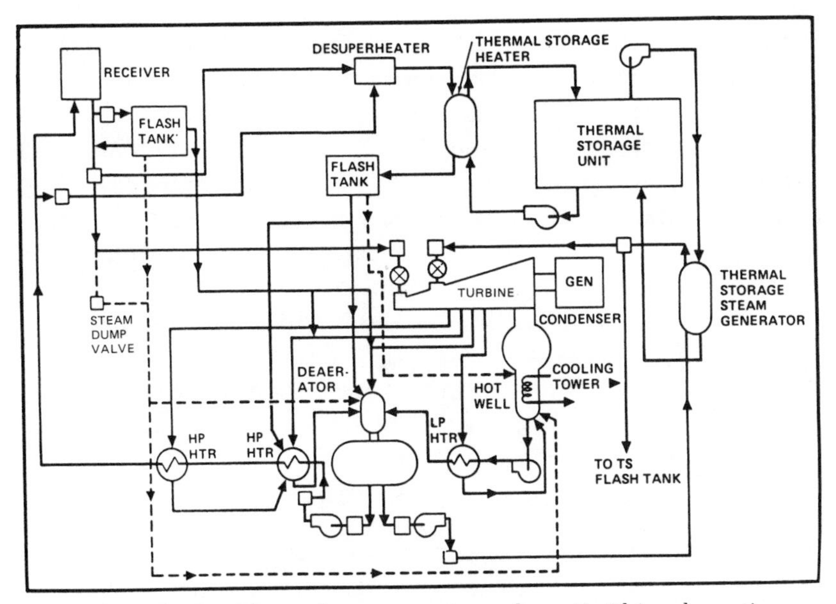

Figure 13-1. Thermal storage system schematic (this schematic is too detailed for the accompanying text).

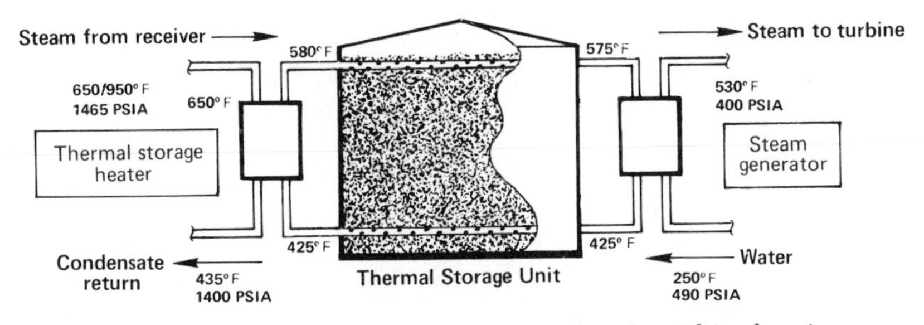

Figure 13-2. Thermal storage system drawing (this drawing matches the precision in the accompanying text).[1]

BEING CLEAR

Too often in scientific writing, illustrations confuse rather than inform readers. Many scientists mistakenly assume that an illustration is automatically worth a thousand words. This is not so. A picture or photograph may raise more questions than it answers. Figure 13-3 supposedly shows a chemical reaction driven by solar energy, but all you can tell from this photograph is that the reaction is bright. This photograph does not focus attention on the experiment; instead, extraneous details stand out. Does this guy always keep his lab this messy? Does he really wear that white lab coat? What is written on the clipboard behind him?

Figure 13-4 is an illustration worth a thousand words. Figure 13-4 is a cutaway drawing of a chamber used to study the electrical breakdown of nitrogen. This drawing efficiently presents many details about the chamber with no lab coats or clipboards vying for attention.

Confusing illustrations typically arise because of careless mistakes made by the writer. A common mistake is making illustrations speak for themselves. Illustrations cannot stand alone; you must use language to introduce illustrations. When you introduce an illustration, you should assign it a formal name such as Figure C or Table VI. Moreover, you should provide enough information in your text so that your readers can understand what the illustration means.

> *A solar central receiver system is shown in Figure 13-5. In this system, a field of sun-tracking mirrors or "heliostats" focuses reflected sunlight onto a solar-paneled boiler or "receiver" mounted on top of a tall tower. Within the receiver, the solar energy heats a transfer fluid that drives a turbine. Electricity is then produced by generators coupled directly to turbines. Most central receiver systems also include a thermal storage system that operates the plant for several hours after sunset or during cloudy weather.*

Another common mistake is unclear titling of figures and tables. You'd be surprised at how many illustrations just appear in scientific writing without titles or with titles so vague that few readers can make heads or tails of the illustrations. What makes for a strong title of an illustration? A strong title

---

Figure 13-3. Weak illustration supposedly showing solar-driven chemical experiment. You can't tell anything from this photograph; another illustration is needed.

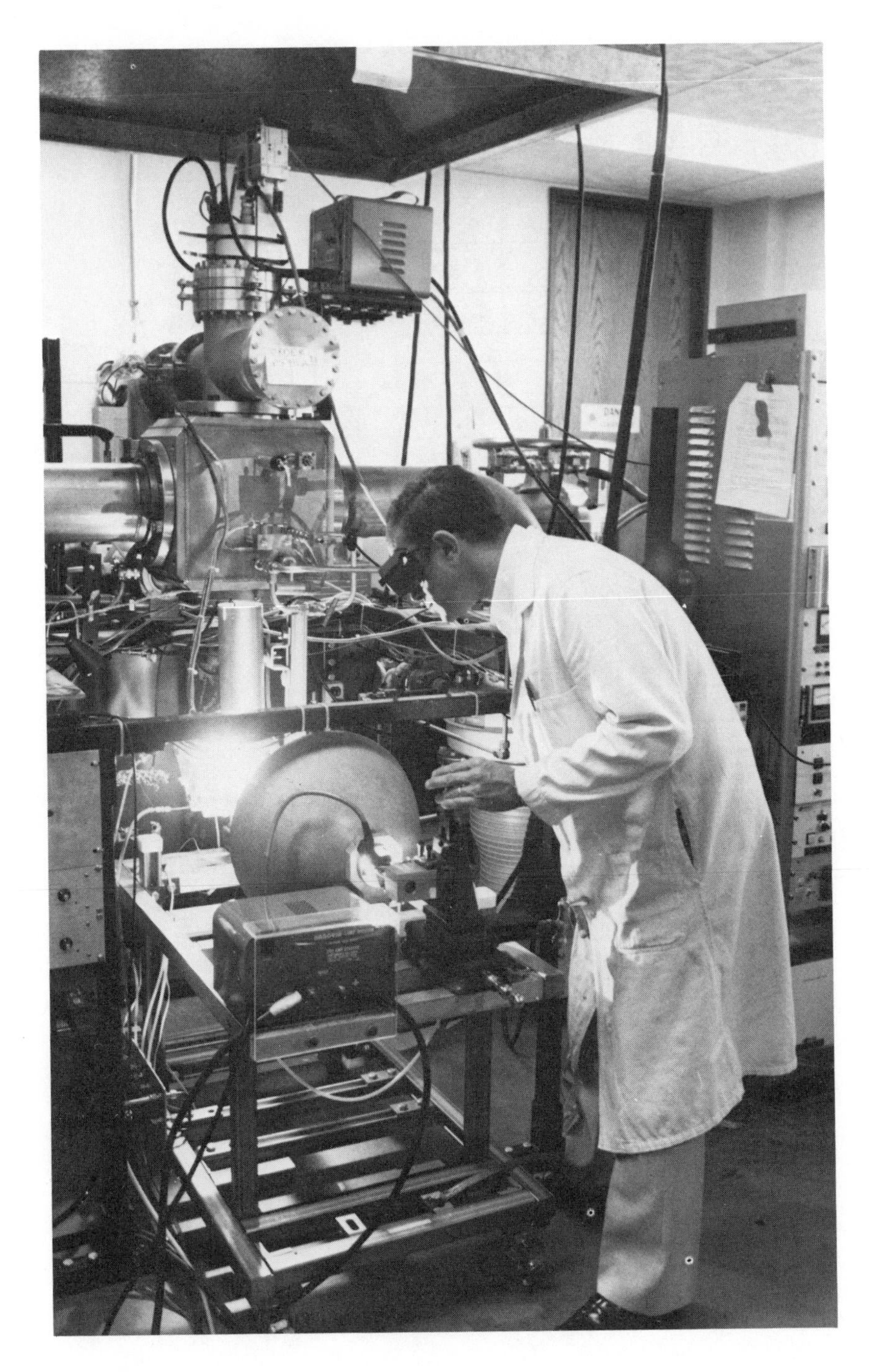

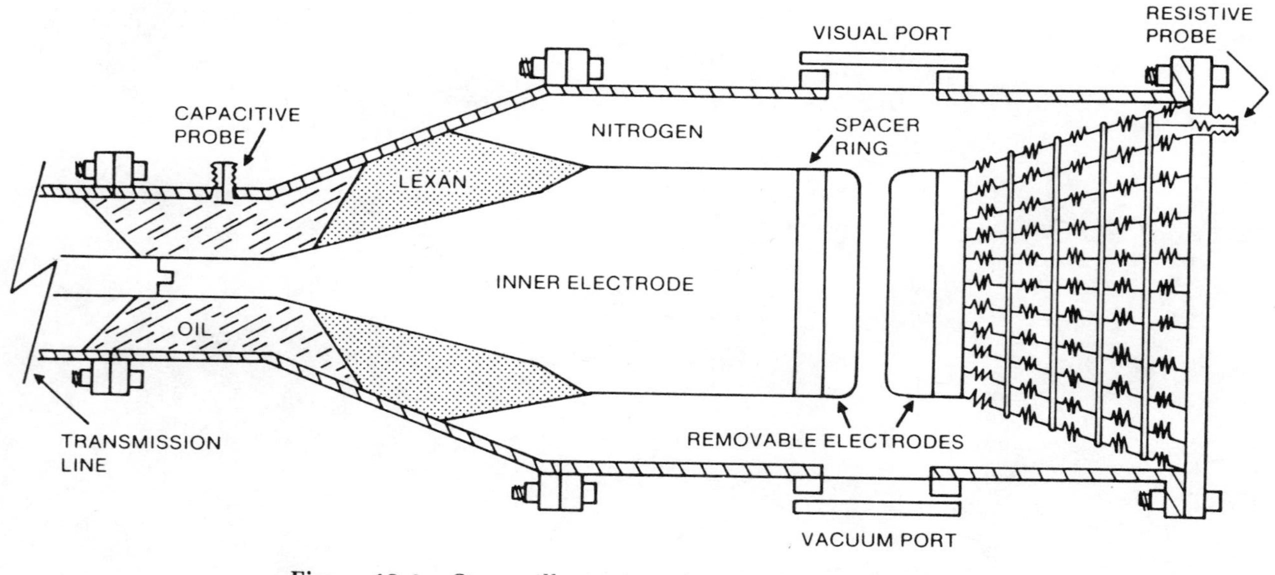

Figure 13-4. Strong illustration of experimental chamber for studying the electrical breakdown of nitrogen. High voltage pulses (~ 100 kilovolts) break down the nitrogen between the electrodes. Capacitive and resistive probes measure the changes in pulse shape and amplitude.[2]

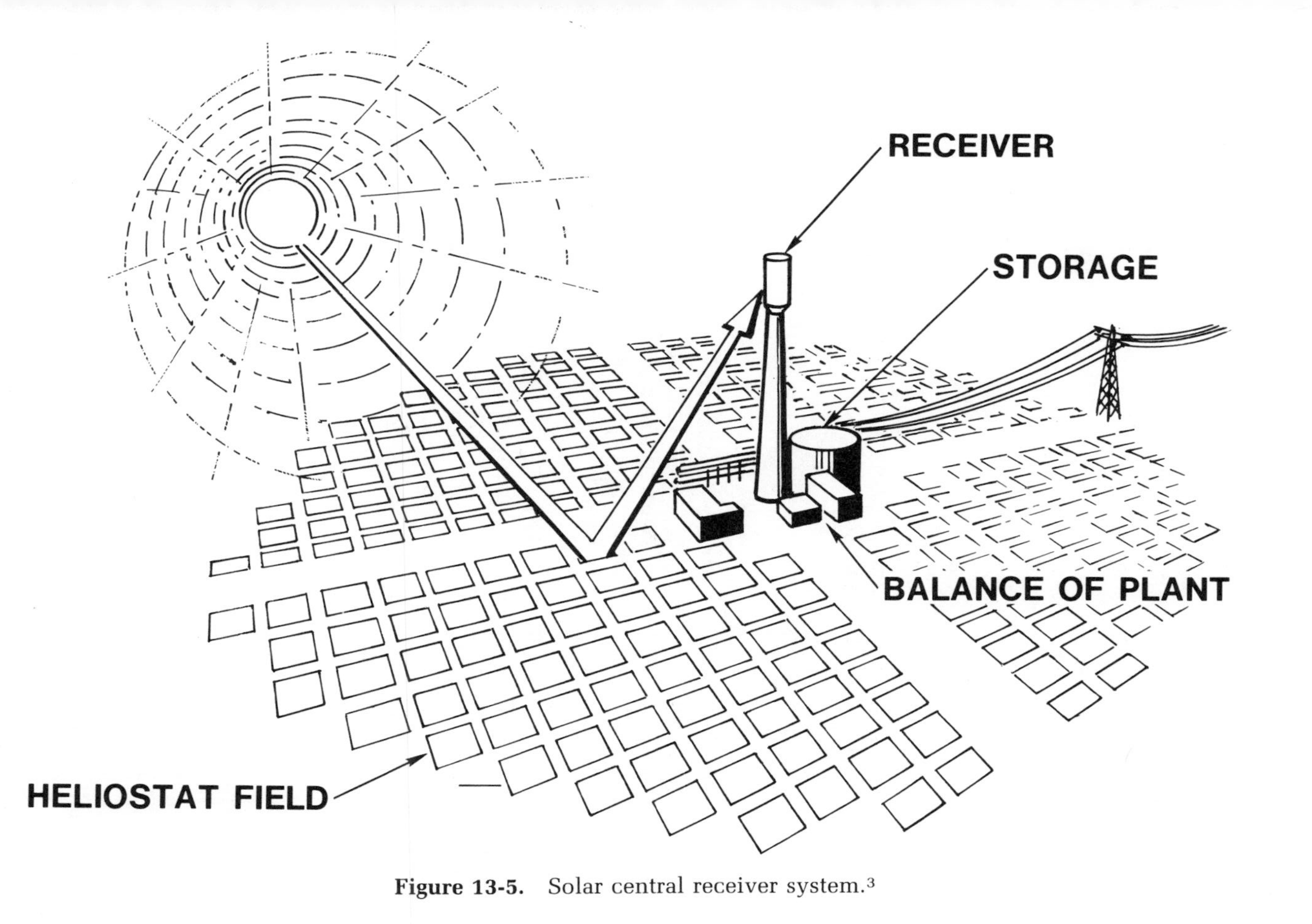

Figure 13-5. Solar central receiver system.[3]

tells readers not only what the illustration is but also separates that particular illustration from all other illustrations in the paper. When readers turn a page, their eyes move first toward your illustrations, then to your illustrations' titles. Therefore, you want your illustration titles to stand independent of the text. A title such as

Figure X. Summary of Data

is too vague. Be more specific:

Figure X. Comparison of predicted and measured heat convection coefficients for a cubical cavity.

However, don't overspecify your illustration:

Table I
Rate Constants k(T) for the Reaction of OH Radicals with C_2H_2; the Indicated Error Limits Are the Estimated Overall Error Limits Which Include the Least-Squares Standard Deviation (2-7 Percent) As Well As the Estimated Accuracy Limits of Other Parameters Such As Pressure and Reactant Concentrations.

This title buries the identity of the table. The chemist should have placed everything after the semicolon in a footnote beneath the table.

Besides having clear titles, you should also clearly label parts within your illustrations. For graphs, you should label horizontal and vertical axes with words (not symbols) and include the units of measurements. For diagrams, you should identify all unusual symbols either in your text or with a key. For photographs, you should clarify important details with call-outs, as shown in Figure 13-6.

BEING FAMILIAR

Anchoring your illustrations in the familiar is important in scientific writing. If your text does not answer questions raised by your illustrations, you will frustrate your readers. To make your illustrations familiar, you must consider your audience: what they know and what they don't know. Because you've probably spent months, even years, with your research, this task is not easy. Nonetheless, you must find a way to see your illustrations as your readers see them—with fresh eyes.

Let's say you wanted to present a sequence of high-speed schlieren photographs as shown in Figure 13-7. Now this sequence of photographs raises several questions for your readers:

Why are the photographs circular?

How large are these circles?

What are the dark spots?

What are the light spots?

What changes (if any) does the sequence show?

Although you may have asked yourself these questions at one time in your research, you might overlook answering them when you write your paper. Don't take chances. In the editing stages of your paper or report, you should show your illustrations to someone who knows little or nothing about your research. Then, listen. You'll be surprised at the questions your illustrations will raise.

One thing you could have easily included in Figure 13-7 was a scale. Something such as

scale ⌐¬ *1 mm*

Too many times, scientists miss the opportunity to show the relative size of something in a photograph, either by inserting a scale or a familiar-sized object such as a penny. Just telling your readers that a heliostat has a surface area of 39 m^2 does not say half as much as showing your readers an illustration such as Figure 13-8.

BEING FORTHRIGHT

To be forthright, you must be straightforward and honest. Unfortunately, many scientists forget this goal when they present their illustrations, particularly their graphs. Presenting graphs is a problem for scientists. On one hand, you want to make readers aware of certain trends. Nevertheless, as a scien-

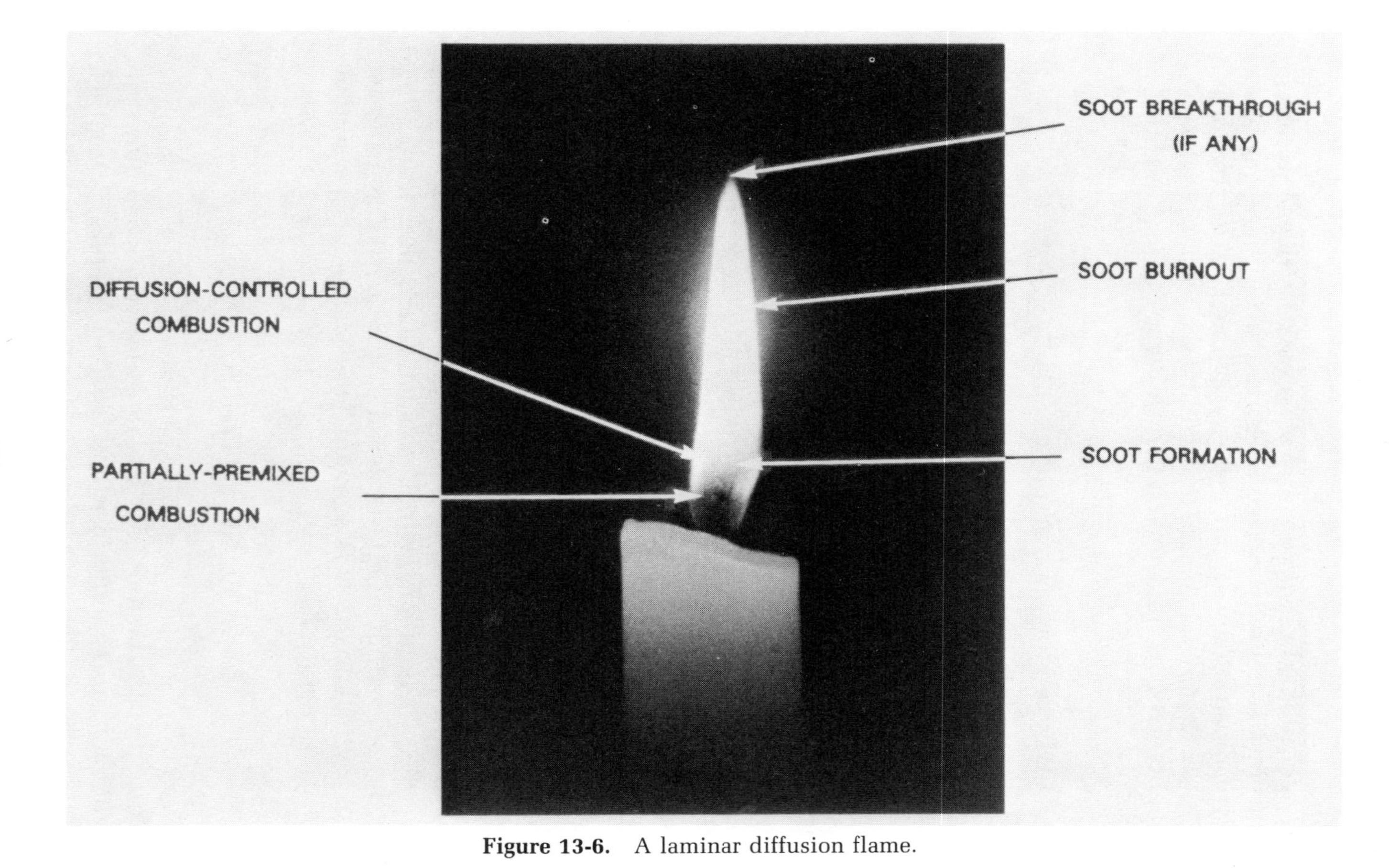

Figure 13-6. A laminar diffusion flame.

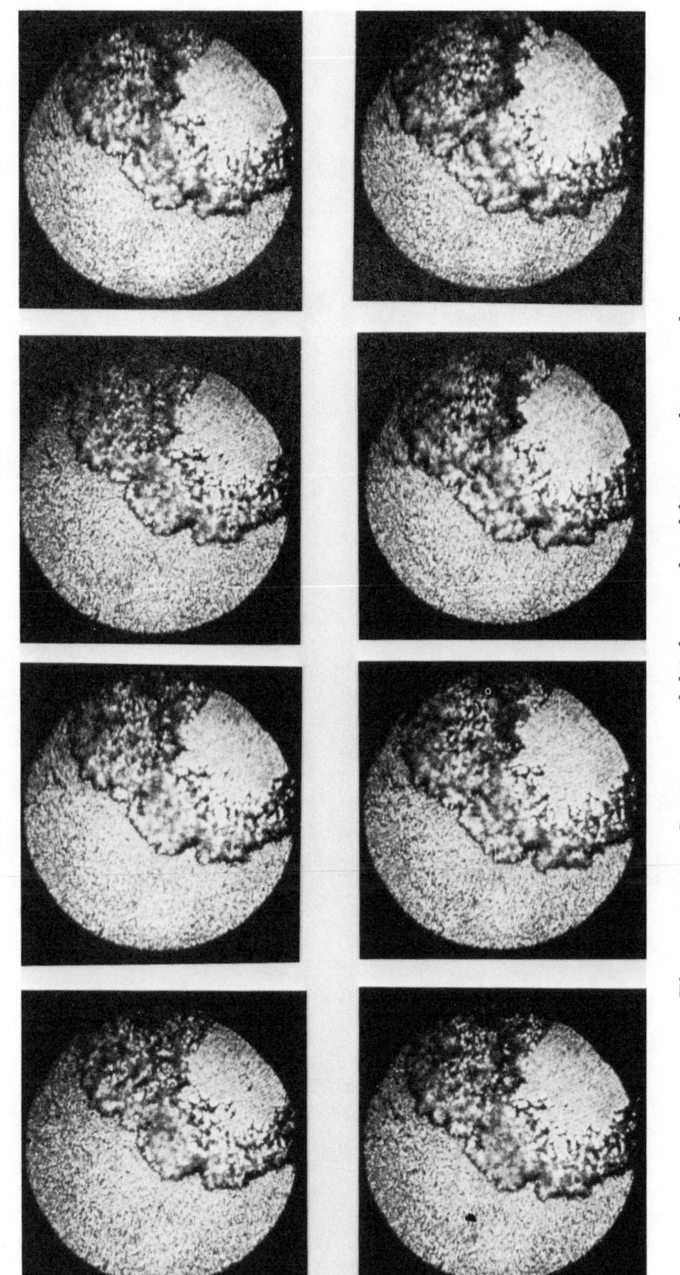

Figure 13-7. Sequence of high-speed schlieren photographs showing how a turbulent flame moves in an engine. The frames are 1.8 μsecond exposures; each exposure is separated by 33.3 μseconds.[4]

Figure 13-8. Heliostats at the Solar One Power Plant.[5]

tist, you have a responsibility not to distort the importance of any trend.

How, then, do you distinguish between a graph that accents and a graph that distorts? This question has no easy answer. In some distorted graphs (such as Figure 13-9a), the horizontal axis does not rest on the vertical zero. You can't just scale all graphs from zero axes. If you did, many valid trends in graphs would disappear; you would bunch all your data points at the top (as shown in Figure 13-9b). What should you do? Well, one thing you can do is give your readers a signal any time your graph's horizontal axis is not at zero. You can signal readers with a statement in the text or with a break in the axis (as in Figure 13-9c).

You can distort graphs in other ways: finagling line fits, deleting data points that don't fit the curve, or else making data points so large that almost any curve would pass through them. Any time you make a graph, you should ask yourself whether your graph brings out a particular relationship or distorts it. If your readers come away from the graph with a disproportionate sense of the relationship, then your illustration is dishonest.

BEING CONCISE

Often in scientific writing, scientists put so much information into their illustrations that readers don't get anything out. Figure 13-10 is a prime example. There's too much ink and not enough white space. The scientist should either delete some of the information, or else use several different graphs to display the information. Granted, the second alternative will not produce a shorter paper, but the paper will be more concise for readers.

Don't think, as many scientists do, that your illustrations have to fill the entire image area of the page. Although you should make illustrations large enough to clarify details, you should also include a reasonable white space border. In scientific writing, white space is important. Papers without white space tire readers.

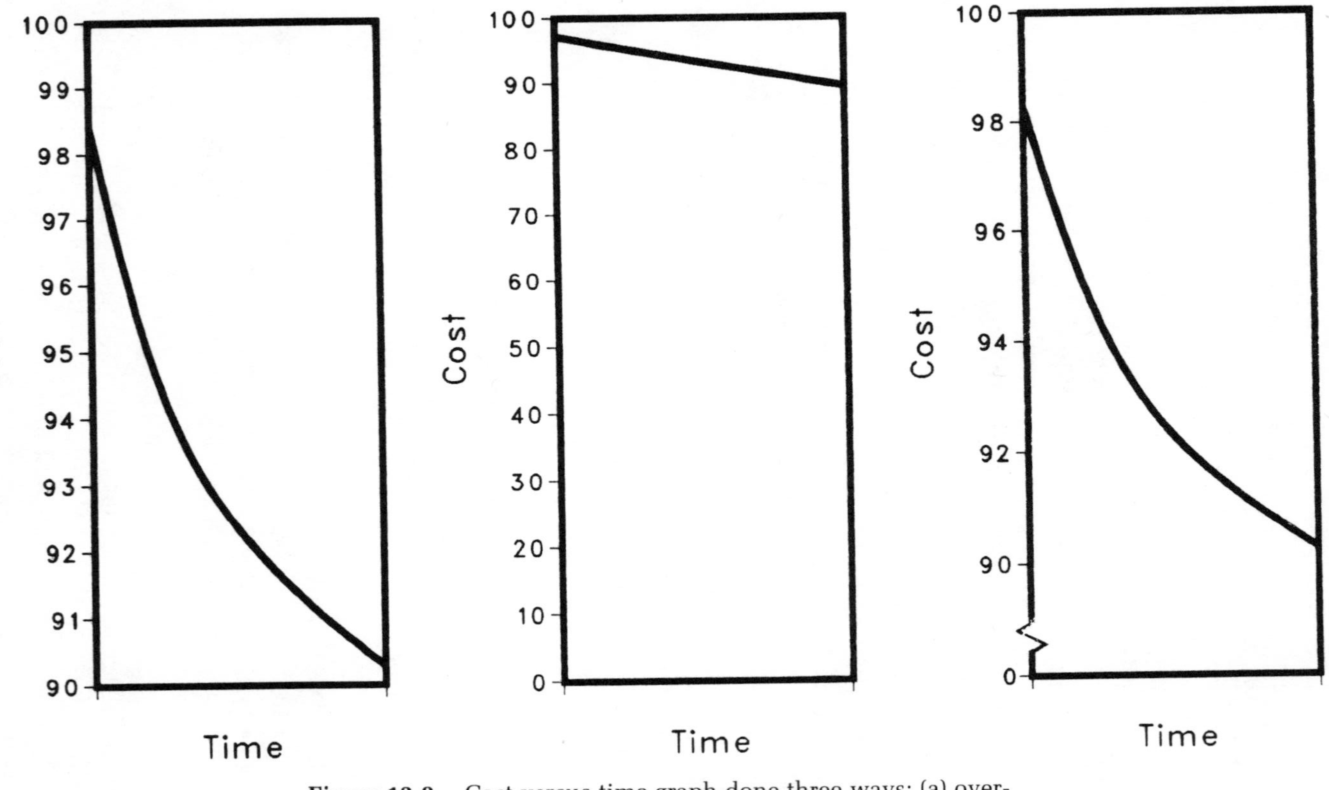

Figure 13-9. Cost versus time graph done three ways: (a) overdramatic case; (b) case in which the scientist buries trend; and (c) case in which the scientist shows trend without overdramatic perspective.

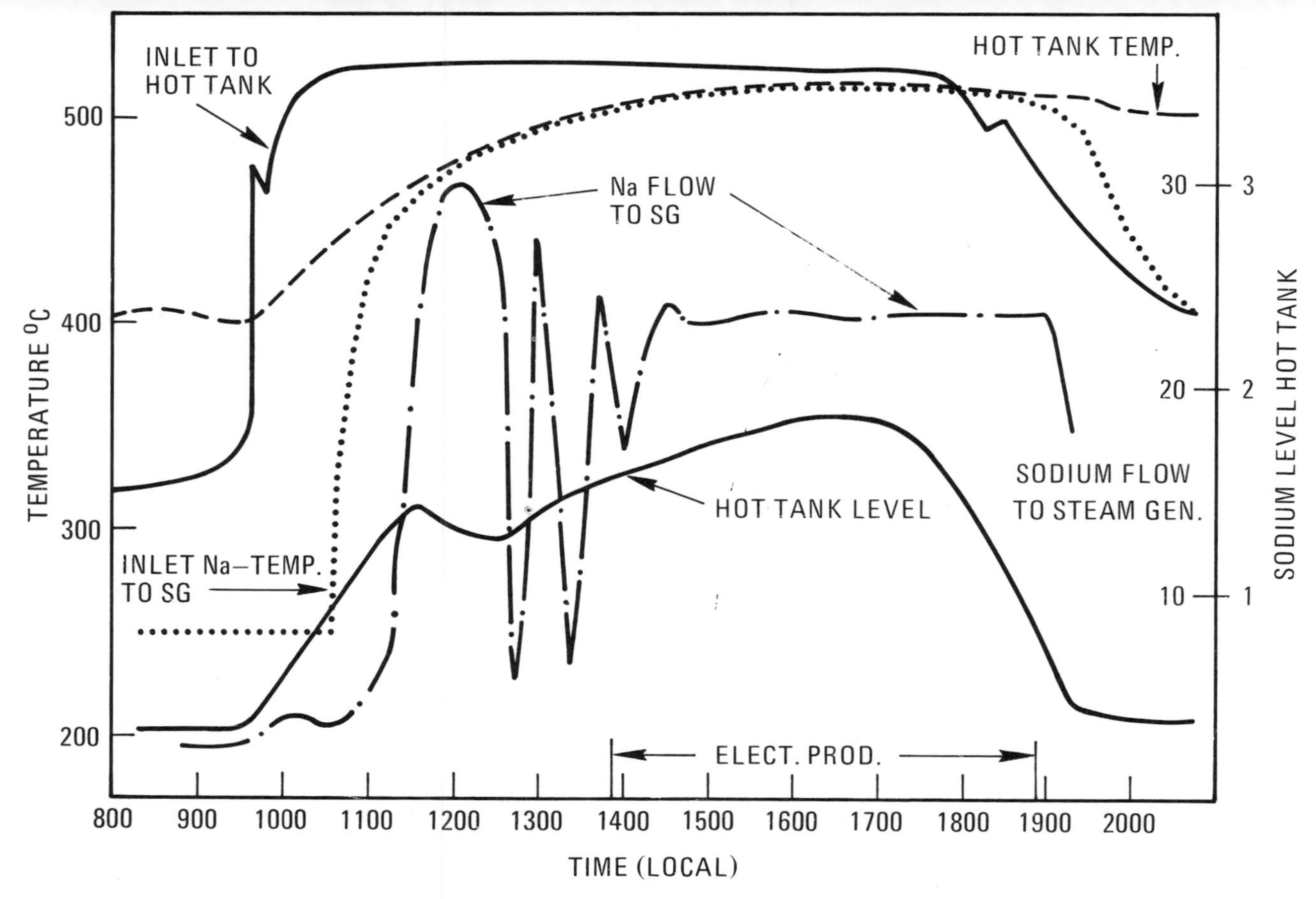

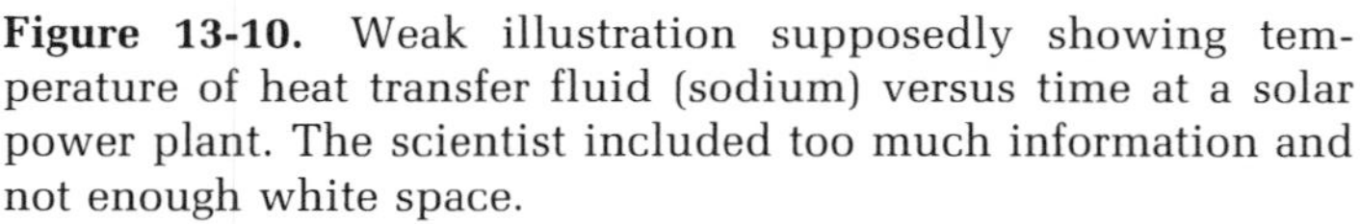

Figure 13-10. Weak illustration supposedly showing temperature of heat transfer fluid (sodium) versus time at a solar power plant. The scientist included too much information and not enough white space.

BEING FLUID

For illustrations to be fluid, you must smooth the transitions between your words and pictures. The most important way to smooth these transitions is to match the information in your text with what's in your illustration. You'd be surprised at how many times scientists will say one thing in their text, then present a figure or table that shows something completely different.

> *The testing hardware of the missile shown in Figure 13-11 has five main components: camera, digitizer, computer, I/O interface, and mechanical interface. Commands are generated by the computer, then passed through the I/O interface to the mechanical interface where the keyboard of the ICU is operated. The display of the ICU is read with a television camera and then digitized. This information is then manipulated by the computer to direct the next command to the I/O interface.*

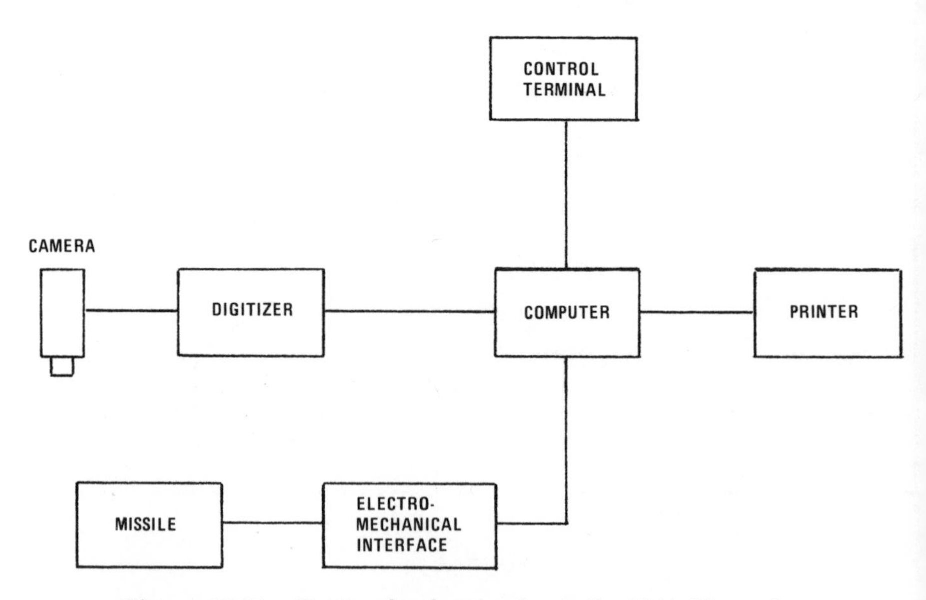

Figure 13-11. Testing hardware of missile (this illustration does not mesh with accompanying text).

This illustration has many weaknesses. First, there aren't five main parts in the figure; there are seven. Two of the parts—the printer and control terminal—are not even mentioned in the text. One of the parts mentioned in the text—the I/O interface (whatever that is)—is not in the figure. What's worse, the scientist blocked the missile in the same manner as the testing system. You can't tell the testing system from the thing that's being tested. In some way, the missile should be set apart from the five main parts of the tester. Finally, you don't gain a sense of the flow of information through this system. The scientist didn't even aim the camera at the missile.

Consider the following revision with attention to our goals for language and illustration:

> *Our system for testing the safety devices of the missile consists of four main parts: computer, camera, digitizer, and electromechanical interface to the missile. In this system (shown in Figure 13-12), the computer generates test commands to the missile through the electromechanical interface. The test results are read with a television camera, then digi-*

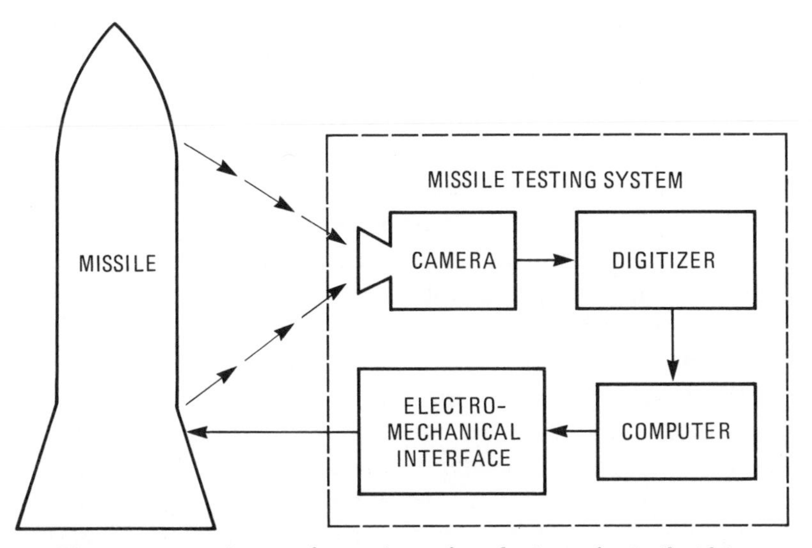

Figure 13-12. System for testing safety devices of missile (this illustration does mesh with accompanying text).

tized. The computer receives the information from the digitizer, then directs the next test command.

To insure smooth transition between your words and pictures, you should (if possible) size illustrations so that they fit vertically on pages facing text. In other words, avoid setting illustrations broadside. You don't want your readers sliding your paper or report all around their desks.

Being fluid also means having illustrations closely follow their text references. Many scientists don't spend enough time laying out illustrations in their papers. Typically, these scientists just place their illustrations at the end of their papers or fit them in wherever there's white space. This method may make things easy for the writer, but it causes difficulty for the readers. Readers have a hard enough time understanding the research without having to wander through a paper to find a particular figure or table.

Time spent laying out your paper is important; as important as the time spent writing your paper.

The worst layout mistake you can make is placing an illustration before its introduction in the text. Readers of scientific papers are not like readers of mystery novels. They rarely read every sentence in order. Even if they do read everything in order, they often skim through certain sections. Therefore, misplaced figures and tables can easily confuse readers. A provocative figure appearing before its introduction in the text will cause readers to read backwards looking for its introduction. Sometimes readers read all the way back to the beginning of the article before realizing (angrily) that the illustration was misplaced. Positioning figures before their introduction in the text is not only unclear writing; it is also inconsiderate writing.

Ideally, you'd like to have your illustrations fall just below the paragraph that introduces them. Unfortunately, page breaks often make this arrangement impossible. The next best arrangement then is to have your illustrations follow closely behind the paragraph that introduces them. This way, readers can quickly compare the text and illustration, and those readers who start with the illustrations and read backwards won't have far to go.

BEING IMAGISTIC

Just having illustrations in your writing does not guarantee that your writing will be imagistic. Typically, only two kinds of illustrations—photographs and drawings—are imagistic; tables, graphs, and diagrams are not. Figure 13-13 is a schematic of a solar power plant. Although this diagram effectively shows how energy passes through the plant, you need a photograph such as Figure 13-14 to show what the plant looks like. You can't say that one of these illustrations is better than the other; that would be the same as comparing apples and oranges. Each type of illustration serves a different purpose.

Furthermore, just having photographs and drawings in your writing does not necessarily mean that your writing will give your readers the strongest images. You must select images carefully. Being imagistic is not the compilation of images, but

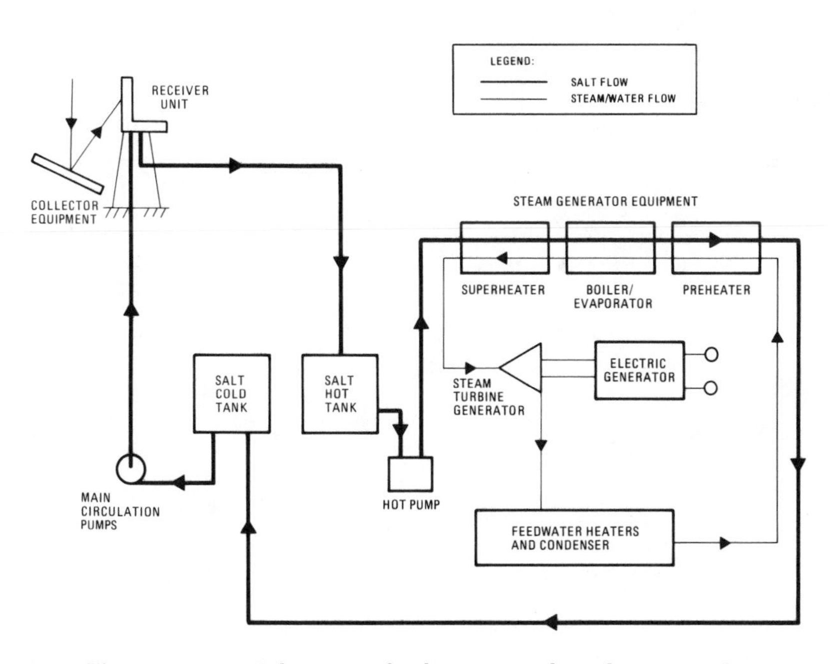

Figure 13-13. Schematic of solar power plant that uses salt as its heat transfer fluid.

the selection of images. For example, if you were writing a paper about the heliostat field of the solar power plant of Figure 13-14, you might want a different perspective for the field—perhaps the one in Figure 13-15. In Figure 13-15, the heliostat field stands out from the other parts of the plant. In a paper about the heliostat field, Figure 13-15 provides the stronger image.

Figure 13-14. Photograph of solar power plant that uses molten salt as its heat transfer fluid. This plant generates 750 kW of electricity, enough for about 250 homes.[6]

Figure 13-15. Photograph of solar plant shown in Figure 13-14. This photograph uses a different perspective, one that accents the heliostat field rather than the tower and piping.[7]

REFERENCES

1. L. G. Radosevich, *Final Report on the Experimental Test and Evaluation Phase of the 10 MWe Solar Thermal Central Receiver Pilot Plant,* SAND85-8015, (Livermore, CA: Sandia National Laboratories, 1986), p. 43.
2. S. J. Levinson, "Investigations of Overvoltage Breakdown," (Unpublished Master's Thesis), Texas Tech University, 1981.
3. Jack Bishop, "Solar Central Receiver System," SL-21085, (Livermore, CA: Sandia National Laboratories, 1986).
4. J. R. Smith, "Turbulent Flame Structure in a Homogeneous-Charge Engine," *Trans. SAE,* 80, no. 820043 (1982), pp. 808–816.
5. A. C. Skinrood and D. N. Tanner, *Solar One—In Perspective,* (Livermore, CA: Sandia National Laboratories, 1986), p. 16.

6. W. R. Delameter and N. E. Bergan, "Molten Salt Electric Experiment," *Solar Central Receiver Bimonthly Report (June-July)*, (Livermore, CA: Sandia National Laboratories, 1985).

7. Ibid.

PART FOUR

Structure

CHAPTER 14

The Strategy of Style

A whole is that which has a beginning, middle, and end.

Aristotle

Developing a structure for a scientific paper involves much more than just organizing the paper into sections. Most scientific papers have the same basic sections:

Summary
Introduction
Discussion of the Research
Conclusion

In scientific writing, structure not only comprises the naming of these sections, but also the way these sections are put together: what information is included, what information is left out, what information is emphasized, what information isn't. Structure is the strategy of scientific writing.

How do you develop a structure? In many types of writing there are set formulas for structure. Gothic romances, for example, always have the first couple of chapters devoted to the heroine: what she looks like, what kind of castle she lives in, what her love life has been like. The next couple of chapters belong to the hero. Then there are a couple of chapters that bring the hero and heroine together, a couple of chapters that tear them apart, and a final chapter to bring them together again.

In scientific writing, though, you cannot use any formulas for structure because every scientific paper has its own beginning, middle, and end. Research is the "search for something new," and you cannot apply a formulaic structure to something that does not follow a set pattern. Structure is the direction of your writing; it is the strategy of your style. Do you lead your readers straight up the face of a sheer cliff? Or do you have them wind along gentle sloping ridges? Structure is the most important element of your style. When your language or illustration falters, your trail breaks and you trip your readers. But when your structure falters, your trail ends and you lose your readers.

There are no simple rules for structure. No cookbook recipes. No packaged outlines. Many scientific writing books approach structure by outlining special strategies for all kinds of writing situations. In these books, you'll learn strategies for informal feasibility reports, formal feasibility reports, straight research reports, special research reports, minor-project reports, progress reports, periodic reports, special-event reports, memo reports, proposals (both external and internal), process descriptions, mechanism descriptions, and on and on.

This approach is silly. For one thing, who wants to memorize all these strategies? For another thing, when applied to real writing situations, these strategies just don't work. Granted, proposals are fundamentally different from feasibility reports; but lumping all proposals into a package and using a single "strategy" for that package is ridiculous. How can *one* proposal strategy work for both a ten-million-dollar grant to contain nuclear fusion with magnetic mirrors as well as a fifty-thousand-dollar grant to study the effects of insecticides on artichokes? If the strategy does work for both projects, then it's so general it's not worth memorizing. If the strategy does not work, as is probably the case, then you've sacrificed grant money for the convenience of a cookbook strategy.

In scientific writing, structure is much too complex to be handled by any write-by-numbers strategy. Structure depends on your particular research and your particular audience, and the person who best knows your particular research and your particular audience is you. Do not think though that finding a strong structure will be the same as taking shots in the dark. Your choices in structure, just as your choices in language and illustration, must again meet the two constraints of scientific writing:

1. You must inform your audience as efficiently as possible.
2. You must be honest.

The first contraint demands that you structure your papers for your readers. Many scientific papers flounder because scientists write down what happened in the order that things happened, rather than in the order easiest for readers to understand. In scientific writing, you must communicate efficiently. An efficient structure does not necessarily mean the shortest paper in length; an efficient structure means the paper that takes your reader the shortest time to understand.

The second constraint of scientific writing demands that you don't structure your paper in a way that distorts your results. For example, you can't bury unfavorable results in the middle of a lengthy discussion. Granted, not all results in a series of experiments or computations should receive the same emphasis. In any given research project, some results will be more important than others. The key word here is "important." Any experimental result that collapses your theory *is* important and demands recognition.

CHAPTER 15

The Beginning

If your research is strong, then everyone should know it. So state what you did up front—the first paragraph, even the first sentence—and leave the dilly-dallying for people who don't have anything to report.

Erich Kunhardt

The beginning to a scientific paper includes not only the paper's introduction, but also its title and summary. The beginning to your scientific paper is a make-or-break situation; it determines whether your audience will read your research. If your beginning is strong, your paper will attract your intended audience. If your beginning is weak, your paper may not attract any audience. In a strong beginning, you state what your research is, you tell what your main results are, and you prepare readers for understanding how you got those results.

THE TITLE

The title is the single most important phrase of a paper or report. The title tells readers what your research is. If your title is inexact or unclear, many people for whom you wrote your paper will never even read it. A strong title orients readers in two ways: (1) it establishes which mountain your research is on; and (2) it separates your research from all other research on your mountain.

A strong title identifies what mountain your research is on. A title such as

Effects of Humidity on the Growth of Avalanches

does not identify the mountain of research. Is this paper a geological study of mountain avalanches in Alaska? Is this paper a study of electron avalanches in electrical gas discharges? This particular paper happened to be the latter. Therefore, a stronger title would have been

Effects of Humidity on the Growth of Electron Avalanches in Electrical Gas Discharges.

In this revised title, you identify the mountain of research.

Strong titles also separate your research from all other research on the mountain. A title such as

Studies on the Electrodeposition of Lead on Copper

is too general. Hundreds of studies fall under this title. If you can concisely specify your research, then do it:

Effects of Rhodamine-B on the Electrodeposition of Lead on Copper.[1]

This title is precise without being unclear; it identifies the mountain of research (surface physics of metals) and separates the research from other research on that mountain. Suppose, though, that the principal aspect of your research was not so much the effect of rhodamine-B on the electroplating process as the use of a new technique—AC-Cyclic Voltammetry—to study organic agents during electroplating. Then your title should read

Use of AC-Cyclic Voltammetry to Study Rhodamine-B in the Electrodeposition of Lead on Copper.

This title emphasizes the unique element of the research: AC-Cyclic Voltammetry.

Titles must not only be precise; they must also be clear. Many times, to be clear in a title, you must cut some details. Some scientists confuse readers by packing too many details in titles. Readers don't know what the principal aspect of the research is.

> Effects of Rhodamine-B and Saccharin on the Electric Double Layer during Nickel Electrodeposition on Platinum Studied by AC-Cyclic voltammetry.

What is the principal aspect of this research that makes it unique? If the principal aspect is the use of the new technique (AC-Cyclic Voltammetry) then a stronger title would be

> *Use of AC-Cyclic Voltammetry to Study Organic Agents in the Electrodeposition of Nickel on Platinum.*

However, if the principal aspect of the research is the effect of rhodamine-B and saccharin on the electric double layer, then you should delete the phrase about AC-Cyclic Voltammetry:

> *Effects of Rhodamine-B and Saccharin on the Electric Double Layer During the Electrodeposition of Nickel on Platinum.*

Ideally, you want your title to identify your research so that it stands apart from any other research on your mountain. Often you cannot achieve this goal in a phrase that is both clear and precise. Nonetheless, you can almost always find a phrase that identifies the most distinctive aspect of your research.

THE SUMMARY

The summary tells readers what happened in your research. In short, the summary gives away the show right from the beginning and allows readers to decide quickly whether they want to read your paper. The summary is the single most important section in a scientific paper. Scientific writing is not mystery writing in which the results are hidden till the end. In scientific writing, you state up front what happened, then use the rest of your paper to explain how it happened.

Many scientists find the principle of summarizing their work at the beginning difficult to swallow. They don't believe that audiences will read their papers all the way through if the results are stated up front. They're right—many readers, after seeing a summary, will not read the entire paper. However, readers who are truly interested in the research will, and the goal of scientific writing is not to entice all audiences to read to the end of your paper, but to inform interested audiences as efficiently as possible. If your research and writing are strong, then interested audiences will continue reading.

Moreover, readers often need summaries to make it through complex scientific papers. Not being told what is going to happen in a scientific paper is the same as being blindfolded and forced to hike a difficult trail. You aren't sure in which direction you're headed or how far you'll be going; and soon after the trail gets rough, you're ready to quit. The same is true for a paper that doesn't state its results up front. You're not sure if it's worth pressing on. For instance, in a paper filled with Monte Carlo simulation techniques, you may tire if you don't know what those simulations really do. If you know that those simulations shed new light on nuclear fusion reactions, then you might stay with the paper. By knowing the paper's destination, you can skim the sections about Monte Carlo techniques without fear of missing the paper's main findings.

In scientific writing, there are two kinds of summaries: descriptive and informative.

The Descriptive Summary

A descriptive summary (often called the *abstract*) tells readers what the research is about; it is like the byline to a baseball game:

California (McCaskill 9-8) versus New York (Niekro 12-9).

From the byline, you know what's going to happen—which teams will play and who will be the pitchers. Descriptive summaries pretty much give the same kind of information about the research, namely what the research will cover.

The first thing a descriptive summary does is identify the research. Don't think that by repeating words from the title in the descriptive summary you are being redundant. A redundancy is a *needless* repetition of words and phrases. Repeating key words and phrases from the title in your summary is important; it corroborates your reader's first impression of your research and gives your reader confidence in the paper. Moreover, in your summary you have room to give details about your research that couldn't fit in the title.

Effects of Ozone and Sulfur Dioxide on Tomato

This paper presents results from a thirty-month study on the effects of urban pollution on tomato plants. The two pollutants studied were ozone and sulfur dioxide, and the area of study was California's Central Valley which harvests 90% of this country's processing tomatoes.[2]

In this summary, you have room to specify the length and location of the study, two details that couldn't fit easily into the title.

A descriptive summary is a table of contents in paragraph form; it is a general map for readers. Descriptive summaries do not present the actual results of the research. Instead, descriptive summaries lay out the central questions that the research will tackle. Because descriptive summaries only state what a scientific paper is going to do, they are typically two or three sentences long. Furthermore, they can usually be written before the paper is written.

This paper will describe a new inertial navigation system for mapping oil and gas wells. We will compare mapping accuracy and speed for this new system against conventional systems.

The Informative Summary

An informative summary (sometimes called the *executive summary*) is the second kind of summary. Unlike descriptive summaries, informative summaries do present the actual result of the research. Informative summaries are analogous to baseball box scores (see Figure 15-1). From a box score, you know what happened in the game: who won, who the winning pitcher was, who had the game winning RBI. Informative summaries give you the same kind of information; namely, what happened in the research.

Like the descriptive summary, the informative summary

YANKEES 4, ANGELS 0

California	ab	r	h	bi	**New York**	ab	r	h	bi
Pettis cf	3	0	1	0	RHndsn cf	3	1	2	0
RJones dh	3	0	0	0	Rndlph 2b	3	1	0	0
Beniqz ph	1	0	0	0	Mtngly 1b	4	0	0	0
Carew 1b	4	0	1	0	Winfld rf	4	1	1	2
Downing lf	4	0	0	0	Hassey c	4	0	2	0
JKHowl 3b	4	0	0	0	Pasqua lf	4	0	0	0
Hendrck rf	4	0	0	0	Baylor dh	2	0	0	0
Wilfong 2b	2	0	0	0	Pgirulo 3b	3	0	1	1
Schofld ss	2	0	0	0	Meachm ss	2	1	1	0
Boone c	3	0	2	0					
Totals	**30**	**0**	**4**	**0**	**Totals**	**29**	**4**	**7**	**3**

California	**000**	**000**	**000 --**	**0**
New York	**300**	**000**	**10x --**	**4**

Game Winning RBI - Winfield (15).

E - R Henderson, McCaskill, Boone. DP-California 1, New York 1. LOB-Cal. 6, New York 7. 2B-Boone. SB-Wilfong (4), RHenderson (58), Pettis (40). S-Meachm.

	IP	H	R	ER	BB	SO
California						
McCaskill L. 9-9	8	7	4	3	4	4
New York						
Niekro W. 13-9	7	4	0	0	3	4
Righetti	2	0	0	0	0	3

Niekro pitched to 1 batter in 8th.

HBP-Baylor by McCaskill. PB-Hassey 2. T-2:43. A-32,189.

Figure 15-1. Box score from Niekro's 297th career victory.

begins by identifying the research. Informative summaries, however, go much further; they state the main results of the research. Informative summaries give the paper's bottom line.

This paper describes a new inertial navigation system that will increase the mapping accuracy of oil wells by a factor of ten. The new system uses three-axis navigation that protects sensors from high-spin rates. The system also processes its information by Kalman filtering (a statistical sampling technique) in an on-site computer. Test results show that the three-

dimensional location accuracy is ±0.1 meter per 100 meters of well depth, an accuracy ten times greater than conventional systems.

Besides mapping accuracy, the inertial navigation system has three other advantages over conventional systems. First, its three-axis navigator requires no cable measurements. Second, probe alignment in the borehole no longer causes an error in displacement. And third, the navigation process is five times faster because the gyroscopes and accelerometers are protected.[3]

This informative summary is tight; there is no needless information. Informative summaries are a sum of the significant points, and only the significant points, of the research. Informative summaries are independent of the paper itself. After reading the informative summary, the reader need only read the paper to find out how the research was done, not what happened.

Everything written in an informative summary—every phrase and illustration—is either a repetition or condensation of something in the paper. Because informative summaries are completely drawn from the paper, they are the last section written. Typically, informative summaries are about 5% of a paper's length.

Which type of summary should you use? Ideally, you'd like to use both. Sometimes, though, formats for papers allow room for only one summary. In these cases, you should choose an informative summary because informative summaries give readers more information.

THE INTRODUCTION

The introduction is the place where readers settle into the "story" of your research. How should you start? You should start strong. Strong introductions to scientific papers don't waste words. More than likely, when your readers begin your introduction, they have specific questions about your research. Your title and summary have stirred their interest and they're anxious to learn more. Don't waste this moment, as so many scientists do, with a first paragraph full of nebulous detail.

> *As is well known, the use of gaseous insulation is becoming increasingly more widespread, with gases such as air and sulphur hexafluoride featuring prominently. There has also been some discussion of using mixtures like nitrogen and sulphur hexafluoride, and of course nitrogen is the major constituent of air.*

What a wimpy beginning. Journalists will tell you that a first sentence is the most important sentence in an article. Make your first sentence count:

> *This article describes what we believe is the first imaging of laser-induced fluorescence from atomic hydrogen. . .*[4]

This chemist didn't waste time. His research was strong and he stated it up front; first sentence, first paragraph. Stating what you did up front gives you a good launching pad. Don't assume though that you have to begin every introduction with a statement about what your research is. The way you begin your introduction depends on your audience. When you write your introduction, you should try to imagine what your readers are thinking. They have just read your title and summary, and even though they have a general idea about your research as well as what happened in it, they probably still have questions:

What *exactly* is your research?

Why was it done?

What do I need to know to understand it?

How will you present it?

Strong introductions answer these questions with clarity and precision. Don't assume that you must address these questions in any order; or that you must give equal space to each question; or that you must address all four questions in every introduction you write. Again, the way you write your introduction depends on your audience and your research. In one paper, you may use most of your introduction telling readers why you did the research. But on another paper, your readers may implicitly know the importance of your research from the title. The important thing is that your readers don't reach the middle of your paper with any of these four questions still burning.

A strong introduction precisely identifies the research. Your introduction is your first chance to fully explain your research. You're not cramped by space as you were in your title and summary. Therefore, you should seize the moment:

> *This paper presents a model to describe the electrical breakdown of a gas. We call this model the two-group model because of the similarity between the problem of gas breakdown and the problem of neutron transport in nuclear reactor physics. The two-group model is based on electron kinetics and applies to a broad range of conditions (breakdown in pure gases, for example). The model also provides a continuous picture of the initial phase of breakdown above the Townsend regime, both in the structure of the breakdown and in the physics of the processes.*[5]

This introduction gives details about the research that couldn't fit in a title or summary, details such as where the theory got its name and its relation to other theories.

When you identify your research in your introduction, you should include the *scope* and *limitations* of the research. The scope includes the things your research will cover; the limitations include the things your research will not cover. Scope and limitations usually go hand in hand. Often, when you identify a research project's scope, you give readers a hint of its limitations.

> *We have measured the breakdown voltage of nitrogen in uniform electric fields for pressures between 1 and 300 torr and with three different electrode surfaces: aluminum, copper, and graphite.*

Because you specified which gas will be studied and under which conditions, you gave readers a clue as to the things that won't be studied. After seeing this summary, your readers wouldn't expect a discussion about the electrical breakdown of helium in a nonuniform field. Sometimes, though, you must specify your limitations:

> *In this paper, we have compared the life expectancies of three different groups of people: heavy alcohol drinkers, light alcohol drinkers, and teetotalers. We have not, however, studied the social or economic makeup of these groups; two elements which could affect life expectancies much more than alcohol intake.*

In this example, you needed to specify your limitations because your limitations raised important questions that your readers probably would not have inferred from the scope.

A strong introduction tells readers why the research is important. Many scientists mistakenly assume that their readers implicitly understand the importance of the research just from the title. These scientists forgo stating the importance of the research and launch right into the nuts and bolts of their paper. The result is that many readers don't finish reading the paper because they have no reason to. Let's face it: reading scientific papers is hard work, taxing work, and readers need incentives. Another reason to show the importance of your research is *money*. Most scientific research depends on outside funding, and before someone will give you money, you must convince them that your work is important. More often than not, that particular someone will be a nonscientist. You must work *hard* to justify your work to nonscientists. You cannot get away with just saying your research is important.

> *This paper presents the effects of laser field statistics on coherent anti-Stokes Raman spectroscopy intensities. The importance of coherent anti-Stokes Raman spectroscopy in studying combustion flames is widely known.*

This introduction convinces your readers of nothing. Don't just tell readers that your research is important. Instead, show readers why it's important.

> *This paper presents a design for a platinum catalytic igniter in lean hydrogen-air mixtures. This igniter has application in light water nuclear reactors. One danger at a light water nuclear reactor is a loss-of-coolant accident. A loss-of-coolant accident can produce large quantities of hydrogen gas when hot water and steam react with zirconium fuel-rod cladding and steel. In a serious accident, the evolution of hydrogen may*

be so rapid that it produces an explosive hydrogen-air mixture in the reactor containment building. This mixture could breach the containment walls allowing radiation to escape. One proposed method to eliminate this danger is to intentionally ignite the hydrogen-air mixture at concentrations below those for which any serious damage might result.[6]

Don't assume that you have to show a practical application for all your research. Much strong research exists for the sole purpose of satisfying scientific curiosity. In such cases, though, you cannot assume that your readers already share your curiosity; you must instill that curiosity. You must raise the same questions that made you curious when you began the research:

In size, density, and composition, Ganymede and Callisto (Jupiter's two largest moons) are near twins: rock-loaded snowballs. These moons are about 5000 kilometers in diameter and contain 75% H_2O by volume. The one observable difference between them is their albedo; Callisto is dark all over, whereas Ganymede has dark patches separated by broad light streaks. This paper discusses how these two similar moons evolved so differently.[7]

How much space should you devote to justifying your research? That depends on your audience. If your principal readers are experts in your field, you may not have to explicitly justify your research at all; your audience might implicitly understand its importance. However, not justifying your research limits your audience. Your audience, in essence, becomes only that group of experts who implicitly understand the importance of your work. Justifying your research increases your audience; and many times, justification requires no more than a paragraph or two.

Some scientists suspect that an apocalyptic collision occurred some 65 million years ago when a huge meteor crashed into the earth and hurled debris into the air. The sky blackened, the sun's warmth was blocked in the upper atmosphere, and the weather at the earth's surface grew chill. Vegetation died and so, according to this theory, did the dinosaurs.

There is an impressive amount of evidence for the hy-

> *pothesis, and it leads to another question. If a meteor could wreak this sort of havoc, could smoke belching from fires started in a nuclear war have a parallel effect, screening out the sun's warmth and cooling the earth enough to hamper agricultural production if not to threaten animal life directly? This paper presents a study of smoke distributions that would follow a nuclear war.*[8]

A strong introduction gives enough background material to help readers understand the paper. Beginnings to scientific papers prepare readers to understand the research. Beginnings to other kinds of writing don't have this goal. Novels, for example, often begin without any preparation at all. Consider Faulkner's opening to *The Sound and the Fury:*

> *Through the fence, between the curling flower spaces, I could see them hitting. They were coming toward where the flag was and I went along the fence. Luster was hunting in the grass by the flower tree. They took the flag, and they were hitting. Then they put the flag back and they went to the table, and he hit and the other hit. Then they went on, and I went along the fence. Luster came away from the flower tree and we went along the fence and they stopped and we stopped and I looked through the fence, while Luster was hunting in the grass.*
>
> *"Here, caddie." He hit. They went away across the pasture. I held to the fence and watched them going away.*
>
> *"Listen at you now," Luster said. "Aint you something, thirty-three years old, going on that way. After I done went all the way to town to buy you that cake. Hush up that moaning. Aint you going to help me find that quarter so I can go to the show tonight."*[9]

This strategy of beginning in the middle of things (*in medias res*) throws readers off-balance. For instance, you probably didn't realize right away that the men behind the fence were playing golf, or that the narrator was retarded. Moreover, you don't figure out until much later in the book why the narrator starts crying: the golf match triggers the memory of his brutal castration. For *The Sound and the Fury,* though, this kind of beginning works. Faulkner wanted his readers to be somewhat confused when they began this novel.

In scientific writing, however, an *in medias res* strategy has no place. In scientific writing, you cannot afford to throw your readers off-balance. Readers have a hard enough time as it is. Because scientific research is so complex, you must ease your readers into the "story" of the research. You must set the stage, and the way you do that is with background information.

What kind of background material should you give? The kind of background material you give depends on your audience—how high they are on your mountain of research. You can't begin every paper at the lowest stratum of science with Euclid and Archimedes. If you did, every paper you wrote would be book-length. Instead, you must start at an elevation of knowledge near your research, yet an elevation that your readers are familiar with. For example, if it were 1913 and you were Niels Bohr writing the theory of the hydrogen atom, you might assume that your readers were familiar with Balmer's equation for wavelength and Coulomb's law of force, but not with Rutherford's nuclear model of the atom which was proposed in 1911. You would then start your paper at an elevation of knowledge somewhere below Rutherford's work.

You must be selective with background material. Space in a scientific paper is valuable. Provide background on only those things that your audience needs to know to understand your research. Scientists often fill their introductions with nebulous historical discussions that prepare readers for nothing.

> *The last decade has seen a rapid development of new techniques for studying the enormously complex phenomena associated with the development of high pressure discharges . . .*

In your writing, you don't have time for this kind of meandering. You must get to the point.

Also, don't assume that all background material in a scientific paper must go into the introduction. In an introduction, you want only background material that pertains to every section of your paper or report. For example, if your report presented three Raman scattering experiments that used one particular laser, you might include background about that laser in the introduction:

The operating characteristics of the neodymium-glass laser used in our experiments are given in Table 15-1. The laser's 1.5 microsecond pulse duration was short enough to freeze material motion in the engine flame while avoiding optical damage caused by the high peak power of very short (nanosecond) pulses. The output of 5 joules provided enough incident intensity so that we could measure flame temperature from a single pulse. The pulse rate of 10 pulses/second permitted rapid accumulation of data, and the narrow line-width (about 0.03 nanometers) provided sufficient resolution.[10]

Table 15-1
Laser Operating Characteristics

Energy per Pulse	5 joules
Pulse Repetition Rate	10 pulses/second
Pulse Duration	1.5 μseconds
Spectral Bandwidth	0.03 nanometers

This background material fits into the introduction because your audience needs it for every section of your report. But if you had used a different laser on each Raman scattering experiment, you would not place this background material in the introduction. You would place the information in the experimental section where that particular laser was used.

When there are tight length restrictions on your writing, you must rely on references to other papers to supply much of your background information. You should be selective with your "background references." Don't bombard your readers, as many scientists do, with a needlessly long list.

The difficult problems in measuring the composition and properties of hot, luminous, particle-laden flows have been receiving increased attention.[1–7]

Are seven background references actually needed here? Does the writer actually expect the reader to look up all seven references to see the increasing attention that hot, luminous, particle-laden flows are receiving? Would not one reference to a well-written review paper have been more helpful? You could justify giving two references if you believed some of your readers might not have access to a particular journal, but not seven.

Why do scientists give so many background references in their introductions? Do they want to impress upon their audience how well-read they are? You can't sacrifice the first constraint of scientific writing because of your own insecurities. When referencing background material, you should choose one or two strong papers (perferably review papers) that will most help your readers.

A strong introduction gives readers a clue as to the organization of the paper. Showing readers the path you're going to take is important. This is not only true in writing, but in all kinds of communication. Anyone who has ever attended a Southern Baptist revival understands this point. In a Southern Baptist revival, the preacher has no time limit. The only saving grace for the congregation is that most Southern Baptist preachers use a three-point sermon. In a three-point sermon, the preacher states in the beginning the three points that he's going cover—say Sin A, Sin B, and Sin C—and then the preacher covers those three sins, one at a time and in the order that he first stated them. Once he's covered all three sins, the sermon is over and you sing the invitation. This mapping out of the sermon's structure is important to a congregation because at any given moment in the sermon you have an idea about how much longer the preacher will be speaking. If the preacher's only on Sin A, you know you've got a while; but if he's on Sin C, you can start thumbing through your hymnal and planning dinner.

Although your paper will probably have a different strategy than a Southern Baptist preacher's sermon, mapping your paper's strategy is important. Placing this map at the end of your introduction often makes for a strong transition to your next section.

> *This report discusses the effects of smoke on the earth's climate following a large-scale nuclear war. First, we present a war scenario involving 10,000 megatons of high-yield weapons. We then introduce assumptions about the amount of*

> *smoke produced from fires, the chemical characteristics of the smoke, and the altitudes at which the smoke is initially injected into the atmosphere. Third, we use computer models to show how the smoke distributes itself in the weeks and months following the war, and finally we discuss how the earth's climate changes as a result of that smoke distribution.*

Once you have given a map of your strategy, you are obligated to stick to it. Nothing makes a congregation more restless than a preacher who promises to talk about only three sins and then covers four.

REFERENCES

1. J. Farmer and R. Muller, "Effect of Rhodamine-B on the Electrodeposition of Lead on Copper," *J. Electrochem. Soc.*, 132, no. 2 (Feb. 1985), p. 313.
2. Lawrence Livermore National Laboratory, "Effects of Ozone and Sulfur Dioxide on Tomato," *Energy and Technology Review*, (Livermore, CA: Lawrence Livermore National Laboratory, July 1985), p. 80.
3. James Kelsey, "Inertial Navigation Techniques Improve Wellbore Survey Accuracy Tenfold," *Sandia Technology*, 7, no. 1 (1983), p. 22.
4. J. Goldsmith and R. Anderson, "Imaging of Atomic Hydrogen in Flames with Two-Step Saturated Fluorescence Detection," *Applied Optics*, 24, no. 5 (1985), p. 607.
5. E. Kunhardt and W. Byszewski, "Development of Overvoltage Breakdown at High Pressure," *Physical Review A*, 21, no. 6 (1980), p. 2069.
6. L. R. Thorne, J. V. Volponi, and W. J. McLean, "Platinum Catalytic Igniters for Lean Hydrogen-Air Mixtures," *Sandia Combustion Research Program Annual Report*, (Livermore, CA: Sandia National Laboratories, 1985), chap. 7, p. 12.
7. Lawrence Livermore National Laboratory, "Exotic Forms of Ice in the Moons of Jupiter," *Energy and Technology Review*, (Livermore, CA: Lawrence Livermore National Laboratory, July 1985), p. 98.
8. Jeff Garberson, "Climate Change: From the 'Greenhouse Effect'

to 'Nuclear Winter,'" *The Quarterly*, (Livermore, CA: Lawrence Livermore National Laboratory, July 1985), p. 98.

9. William Faulkner, *The Sound and the Fury*, (New York, N.Y.: Random House, Inc., 1946), p. 1.
10. C. B. Layne, "Sharing Laser Beams in a Multi-Lab Research Complex," *Laser Focus*, 17 (1981), p. 83.

CHAPTER 16

The Middle

If a man can group his ideas, he is a good writer.

Robert Louis Stevenson

The middle (or discussion) of a scientific paper presents your research. In your discussion, you state what happened as well as how it happened. You state your results, show where they came from, and explain what they mean. Many scientists mistakenly wait until their conclusion section before presenting any results. These scientists make their readers wait too long. You should treat each result (or set of results) individually in your discussion, then use your conclusion to treat all results in the context of the entire research.

DEVELOPING A STRATEGY

The biggest task in writing the discussion is choosing a strategy; a path for parceling information to your audience. Many scientists fail to find an effective strategy in their writing. Instead of parceling information, these scientists bury readers under avalanches of details:

> *To separate the hot and cold oil, one tank was used that took advantage of the thermocline principle which uses the rock and sand bed and the variation of oil intensity with tem-*

> *perature (8% decrease in density over the range of operating temperatures) to overcome natural convection between the hot and cold regions.*

Although you often feel compelled to tell readers about ten details at once, you must hold back. Readers can't digest that many details in one lump. You must develop a strategy for parceling details in digestible portions.

> *The question was how to separate the hot and cold oil in the rock and sand bed. Rather than have one tank hold hot oil and another tank hold cold oil, we used a single tank for both. This design took advantage of the variation of oil density with temperature. In our storage system, oil decreases in density by 8% over the range of operating temperatures. This variation in density allows the hot oil to float over the cold oil in the same tank. Natural convection is stymied by the position of hot over cold and by the rock and sand bed. Thus, the heat transfer between hot and cold regions is small because it occurs largely by conduction. The concept of storing heat in a single vessel with hot floating over cold is known as "thermocline storage."*

Strategies for parceling information are not like recipes. You can't pull out an appetizer strategy for proposals or a main course strategy for research reports. For each paper, you must develop your own particular strategy. Developing a strong strategy is typically a trial-and-error process. You envision a path, you try it, and then you look back to see if it works for your research and audience. To envision a path, you must know your research and audience. You must imagine your readers starting at some point A on your research mountain and hiking to point B, your final results. To perform the second step in the process (trying a path), you need the energy to outline the path on paper. To perform the third step in the process (checking to see if the path works), you need some critical ability. You must know how to distinguish between a strong and weak strategy.

How do you recognize a strong strategy? This question is rather difficult. Perhaps an easier question is how do you recognize a weak strategy? Most weak strategies have a common trait; they are illogical.

Solar thermal technology concentrates solar radiation by means of tracking mirrors or lenses onto a receiver where the solar energy is absorbed as heat and converted into electricity or incorporated into products as process heat. The two primary solar thermal technologies, central receivers and distributed receivers, employ various point and line-focus optics to concentrate sunlight. Current central receiver systems use fields of heliostats (two-axis tracking mirrors) to focus the sun's radiant energy onto a single tower-mounted receiver. Parabolic dishes up to 17 meters in diameter track the sun in two axes and use mirrors or Fresnel lenses to focus radiant energy onto a receiver. Troughs and bowls are line-focus tracking reflectors that concentrate sunlight onto receiver tubes along their focal lines. Concentrating collector modules can be used alone or in a multi-module system. The concentrated radiant energy absorbed by the solar thermal receiver is transported to the conversion process by a circulating working fluid. Receiver temperatures range from 100°C in low-temperature troughs to over 1500°C in dish and central receiver systems.[1]

The path breaks down here. The first three sentences set up a strategy, but the rest of the paragraph ignores it. The first sentence quickly takes us through a generic solar thermal system—that's okay. The second sentence then breaks down solar thermal systems into two groups: central receivers and distributed receivers. So far, so good. The third sentence then discusses central receivers, but the fourth sentence jumps to parabolic dishes, a term we were unprepared for. The rest of the paragraph exacerbates this mistake by jumping from one detail to another in a Brownian motion. Why didn't the engineer tell us that there were three types of distributed receivers: parabolic dishes, troughs, and bowls? Why were working fluids discussed at the end of the paragraph? Why were certain details made specific (the diameter of parabolic dishes) while other details were left generic (the height of a central receiver tower)?

In strong scientific writing, you don't make these blunders. In strong scientific writing, either you find a logical sequence to lead your readers through the research or else you break your research down into logical parts that your readers can digest. The key word is "logic."

In a solar thermal system, mirrors or lenses focus sunlight onto a receiver where a working fluid absorbs the solar energy as heat. The system then converts the energy into electricity or uses it as process heat. There are two kinds of solar thermal systems: central receiver systems and distributed receiver systems. A central receiver system uses a field of heliostats (two-axis tracking mirrors) to focus the sun's radiant energy onto a receiver mounted on a tower. A distributed receiver system uses three types of optical arrangements—parabolic troughs, parabolic dishes, and hemispherical bowls—to focus sunlight onto either a line or point receiver. Distributed receivers may either stand alone or be grouped.

Finding a Logical Sequence

In scientific writing, logical sequences may be based on time, space, or any number of variables. Which variable you choose depends on your research and your audience. For instance, let's say you studied the combustion of a coal-water slurry, and let's say that you found that the combustion process occurred in eight different stages, as shown in Figure 16-1. You could then structure your discussion in eight steps by following the chronological sequence of the photographs.

If your research was the nuclear fusion experiment in Figure 16-2, you might try a spatial sequence. In this sequence, you would discuss the experiment's parts in an order that paralleled their physical arrangement. For instance, you might begin with the Marx generators on the circumference and move radially inward, finishing with the deuterium-tritium chamber in the center.

For this particular experiment, you could simplify things by choosing a different sequence, say one that followed the flow of energy depicted in Figure 16-3. Using this energy schematic, you would discuss events rather than parts:

Charging Marx Generator

Forming Line Pulse

Generating Particle Beam

Transporting Particle Beam

Irradiating Deuterium-Tritium Pellets

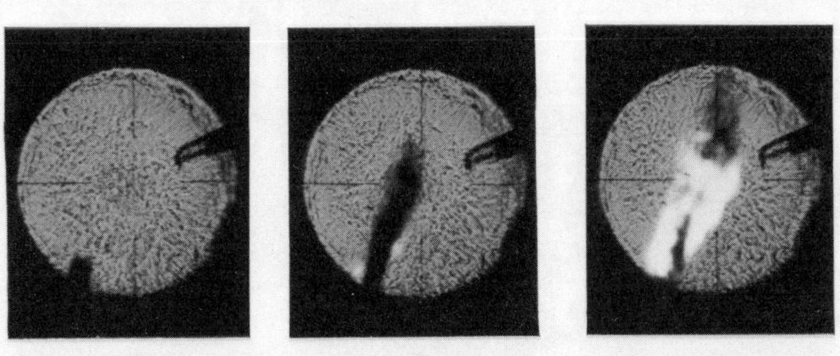

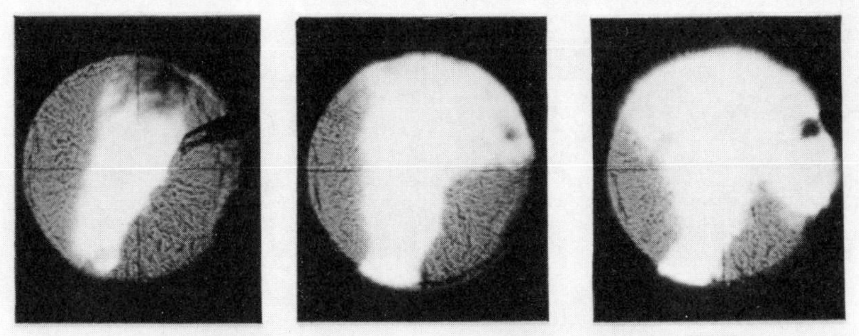

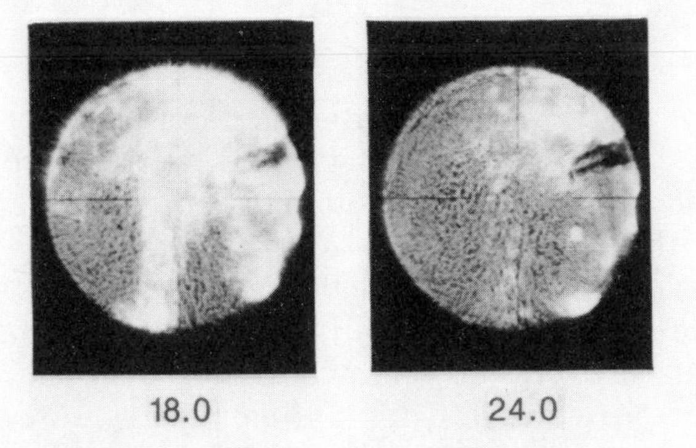

Figure 16-1. High-speed photographs of coal-water slurry igniting and burning under diesel engine conditions (time in milliseconds).[2]

Figure 16-2. Photograph of nuclear fusion experiment. In the experiment, an accelerator produces beams of lithium ions which converge onto deuterium-tritium pellets in an attempt to produce nuclear fusion.[3]

Depending on your audience, you could reverse the order and begin with irradiating the pellets. This reversed order might work better with readers who understood fusion but weren't familiar with particle beams or Marx generators.

Now you could argue that the energy flow sequence through the accelerator is actually chronological. You're right; all five of these events do occur chronologically (although the time scales are several trillion times apart). However, the name you give a sequence isn't important. What is important is the logic of a sequence. In the next example, the sequence defies classification. Is it spatial? Is it chronological? The answer is that no one knows or really cares. The important thing is that the sequence works.

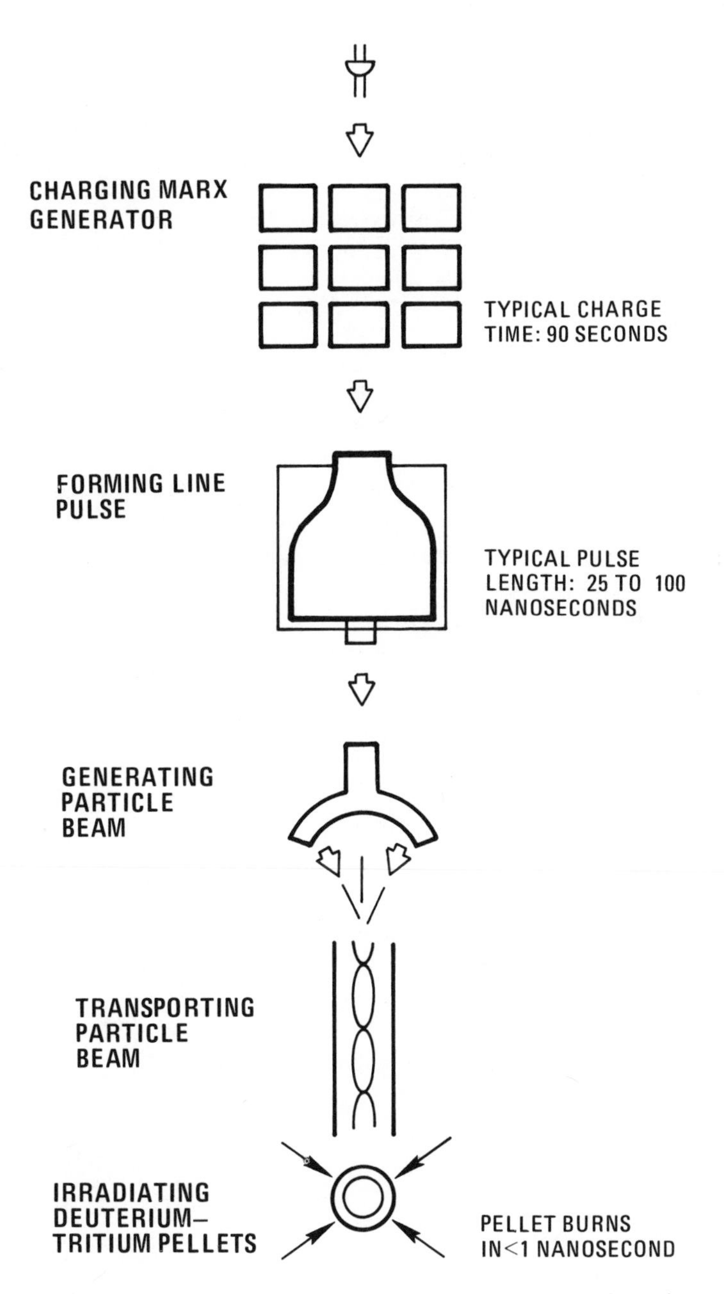

Figure 16-3. Energy flow diagram through nuclear fusion experiment.

A sketch of an electron avalanche moving across a uniform electric field is shown in Figure 16-4. The avalanche shape shows the processes at play in the gas. As depicted, the avalanche breaks down into three initial stages. This division is strictly for illustrative purposes—in reality, a continuous transformation takes place.

In Stage I, diffusion processes determine the radial dimensions of the avalanche. The avalanche radius r_d is given by

$$r_d = (6Dt)^{\frac{1}{2}},$$

where D is the electron diffusion coefficient and t is time. In the regime of interest (where voltages are greater than 20% of self-breakdown) the time of development is so short that little expansion occurs. For the case of nitrogen at atmospheric pressure, $D \simeq \frac{862\text{cm}^2}{\text{sec}}$, $t \sim 10^{-8}$ sec, and $r_d \sim 7.2 \times 10^{-3}$ cm.

As the number of electrons in the avalanche increases, electrostatic repulsion begins to play a role in the expansion of the avalanche. In this stage (Stage II), the avalanche increases exponentially with time. Taking electrostatic repulsion into account, we estimate the avalanche radius as

$$r_e = \left(\frac{3e}{4\pi e_o E_o}\right)^{\frac{1}{3}} e^{\alpha\frac{z}{3}},$$

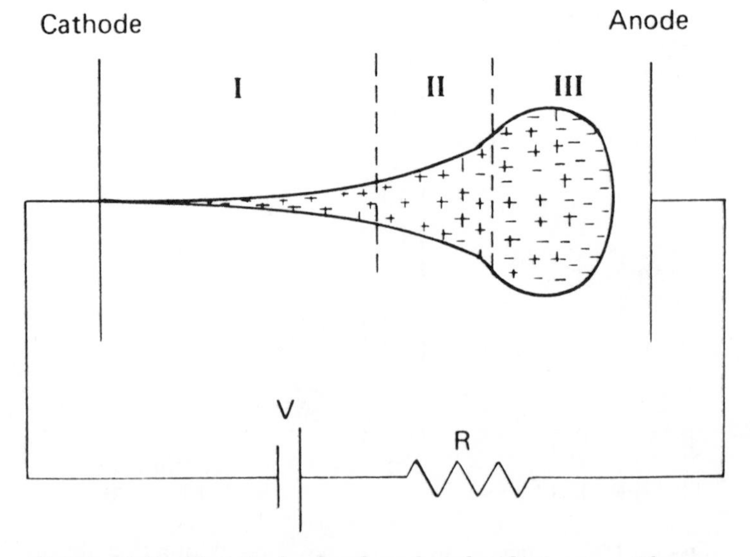

Figure 16-4. Sketch of a developed avalanche in a gas. The gas has been placed under a high voltage.

where e is the electron charge, E_o is the external electric field, α is the Townsend primary ionization coefficient, ϵ_o is the permittivity of free space, and z is the average position along the gap axis of the avalanche with z = 0 being the cathode position. For breakdown in nitrogen at atmospheric pressure and an applied field of $40\frac{kV}{cm}$(which corresponds to a voltage ~25% above self-breakdown), the avalanche radius r_e is 0.14 cm at z = 0.05 cm.

In Stage III, space-charge neutralization begins to take place, and the radial expansion experienced in Stage II slows. In other words, the space charges arising from ionization are distributed in such a way as to slow the radial expansion of the avalanche.

Throughout these three initial stages, as the electron and ion densities in the avalanche increase, two things occur: (1) a highly nonequilibrium electron energy distribution evolves; and (2) the external field distorts because of the space charge. The two effects play a major role in the intermediate stage that follows—Stage IV . . . .[4]

Breaking Research Down into Logical Parts

Many times, you can't express your research in a logical sequence because a sequential path just doesn't exist. In such cases, you should try to break your research down into logical parts. Suppose that you studied the global effects of a nuclear war:

Effects of Radiative Fallout

Effects of Nitrogen Oxides

Effects of Dust

Effects of Smoke

Now you could choose a chronological strategy and discuss each of these four effects during the first week after the war, then the second week, then the third. This strategy proves cumbersome because the effects are so different. A better strategy involves treating each effect separately. Granted, in each individual effect's section, you would probably use a chronological sequence, but your overall strategy would remain to break down the material into logical parts.

What about the way you put parts in order? Which parts should you talk about first? You should choose an order that will most help your readers understand your research. If your parts depend on each other, you should choose the order easiest for your audience to understand. For instance, if you introduced a new nuclear force theory—call it the Q-State Theory—you might break your paper down as follows:

Assumptions of the Q-State Theory

Physical Interpretation of the Q-State Theory

Mathematical Formulation of the Q-State Theory

Experimental Evidence for the Q-State Theory

This breakdown serves its audience well. First, it states the theory's assumptions so that there's no misunderstanding about when this theory applies. Second, this breakdown gives readers a physical picture of what's happening before discussing abstract calculations. This way, equations are anchored with physical images and analogies. Finally, this breakdown corroborates your new theory with experimental evidence.

What should you do when your parts are independent of each other? If your parts are independent of each other, then you should order them according to their importance. In a short paper, it probably makes little difference whether you go in descending or ascending order of importance. Going back to the global effects of a nuclear war, if you found that radiative fallout was the most important effect, you should either discuss it first or last. You might choose an ascending order of importance if you could quickly dispense with the smaller effects; say dust and nitrogen oxides. In this case, your order would be

Effects of Dust

Effects of Nitrogen Oxides

Effects of Smoke

Effects of Radiative Fallout

If your paper is long, however, you should order your effects in descending order of importance. In other words, you should discuss "Effects of Radiative Fallout" first. This way, you don't bury your most important results at the end of a long paper.

How do you determine which part of your research is most important? This decision depends on your audience. When you consider your audience, you must not only know who your audience is, but also why they are reading your paper. Let's say your research was the design of an electronic implant that delivered insulin to a diabetic,[5] and that your research broke down into three parts:

Electronic Design of Implant

Surgical Procedure for Implant

Implant Success in Diabetics

If your audience was a group of doctors, then "Surgical Procedure for Implant" and "Implant Success in Diabetics" would be your two most important sections with "Electronic Design of Implant" a distant third. Doctors would care more about how the device works in people than whether the device used analog or digital circuits. If your readers were electronic engineers, the relative importance of these three sections would change. "Electronic Design of Implant" would probably become your most important section.

MAKING SUBSECTIONS

When you've developed a strategy, you should show your readers what that strategy is by using subsections. It is important to break your discussion into subsections. Not only do subsections allow readers to quickly find information they're most interested in; subsections also allow readers to skip over information they're not interested in. Moreover, subsections (and sub-subsections) give readers white space. Readers of scientific papers need white space to rest and reflect on what was said.

Titling Subsections

How should you title subsections? When you title subsections, you should strive for precision and clarity. For example, if your paper compares a coal-water slurry with dry pulverized coal, you shouldn't title your subsections as

Slurry
  Combustion
  Pollution
Dry
  Combustion
  Pollution

These titles are too vague. Readers often skip around in papers looking for particular results; therefore, you want your titles as clear and precise as possible.

Coal-Water Slurry
  Combustion Efficiency
  Combustion Emissions
Dry Pulverized Coal
  Combustion Efficiency
  Combustion Emissions

When titling subsections, you should also choose parallel titles. For example, in a paper comparing the costs of using a coal-water slurry and pulverized coal, you wouldn't want your subsections broken down as

Transport Cost of Coal-Water Slurry
Combustion Cost of Coal-Water Slurry
Emission Cost of Coal-Water Slurry
Dry Pulverized Coal Costs

The last subsection heading does not parallel the first three. This discussion should either be broken down by fuel (coal-water slurry, dry pulverized coal) or by costs (transport, combustion, emission). For example:

Fuel Breakdown	Cost Breakdown
Coal-Water Slurry	Transport Cost
Transport Cost	Coal-Water Slurry
Combustion Cost	Dry Pulverized Coal
Emission Cost	Combustion Cost
Dry Pulverized Coal	Coal-Water Slurry
Transport Cost	Dry Pulverized Coal
Combustion Cost	Emission Cost
Emission Cost	Coal-Water Slurry
	Dry Pulverized Coal

To see if your sections are parallel, write them down as a table of contents and check the phrase structures. Don't mix verb phrases with noun phrases:

Nonparallel	Parallel	Parallel
Circulator	Circulator	Circulating
Recirculating	Recirculator	Recirculating
Superheater	Superheater	Superheating
Evaporating	Evaporator	Evaporating

Finally, once you title a subsection, don't begin that subsection informally. If your subsection title is "Superheater," don't begin by saying

> *This is responsible for heating the steam above the saturation temperature . . .*

You will throw your readers off-balance. Your readers won't know if you are referring to the superheater or to something else. Instead, adopt a more formal beginning to your subsection:

> *The superheater is the part of the steam generator that heats the steam above its saturation temperature . . .*

Determining Subsection Length

How long should a subsection be? As with most questions about style, there is no absolute answer. Once you decide on a strategy for your discussion, your research and audience should dictate your subsection's length. Don't assume that all subsections must have the same length. Subsection length should parallel subsection importance.

When should you use sub-subsections? Again, this question has no absolute answer. On one hand, you don't want your paper to be a collage of titles and subtitles. On the other hand, you don't want your subsections so long that your readers can't read them through in one sitting. You should figure that your readers will tire after about ten paragraphs. Therefore, if a subsection runs much longer than ten paragraphs and easily breaks down into two or three subsubsections, then break it down. Don't make one sub-subsection if you're not going to have a second.

Effects of Radiation
  Temperature Changes
Effects of Smoke
  Temperature Changes

This breakdown is inherently unparallel: "Temperature Changes" has nothing to be parallel with. You should either include another sub-subsection title, such as "Other Climate Changes," or (if there are no other climate changes) combine the sub-subsection title into the subsection title:

Temperature Changes Due to Radiation
Temperature Changes Due to Smoke

PRESENTING RESULTS

The purpose of scientific writing is to inform your audience as efficiently as possible, and the most efficient way to inform is to answer questions in your text as soon as you raise those questions. Waiting until page 213 of your report to give a result you could have stated on page 16 exasperates your readers. You should state the results of your research as they naturally arise in the discussion. You should not only state your results, but you should also explain what they mean:

> *The predicted convection coefficient $h_{predicted}$ is plotted against the measured convection coefficient $h_{measured}$ in Figure 16-5. The solid line indicates where perfect agreement between the measured and predicted quantities would be. The predicted convection coefficient overpredicts the measured convection coefficient by roughly ten percent. Nonetheless, considering the uncertainties associated with both the data and the prediction, the agreement between $h_{measured}$ and $h_{predicted}$ is quite good.*[6]

Notice that this engineer didn't leave her readers wondering about the discrepancy between the predicted and measured coefficients. She addressed the question as soon as it arose in her discussion instead of waiting until her conclusion section.

When you are ready to give important results to your readers, you should accent those results. There are many ways to accent results: stylistic repetition, illustration, language.

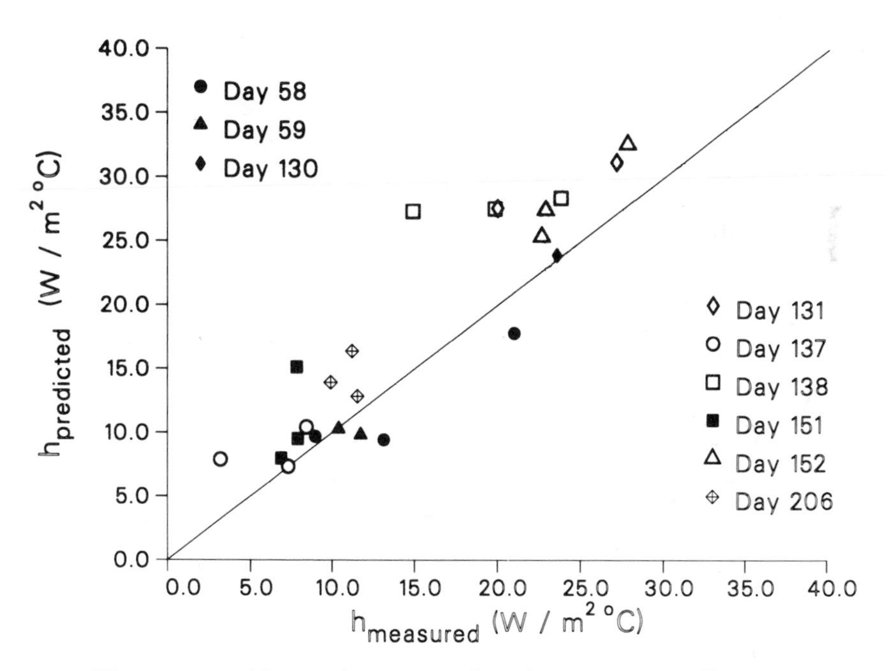

Figure 16-5. Measured versus predicted convection coefficient of the Solar One receiver.

Accenting Results by Stylistic Repetition

If you mention a particular result in your summary, a second time in your discussion, and then a third time in your conclusion, your readers will realize that result is important. This kind of repetition is not being redundant. You are being redundant when you repeat something unimportant. When you repeat something important (and your results are the most important part of your research) it is considered stylistic repetition. Next time you watch a commercial on television, count how many times the commercial repeats a product's name. Don't be surprised if the commercial repeats the name seven or eight times. Advertisers know the value of repetition.

You must use stylistic repetition selectively, though. If you stylistically repeat too many results, you dilute the importance given by repetition.

Accenting Results with Illustration

Another way to accent a result is through illustration. In a paper or report, scientific readers don't always read every sentence. Instead, they skim through parts they think are unimportant. Scientific readers almost always look at every illustration though. Therefore, if you can place important results in an illustration, do so. Illustrations, such as Figure 16-6, stand out in reports and papers.

As with stylistic repetition, a large number of illustrations dilutes the importance given to any one. If Figure 16-6 was one of thirty such global illustrations, it wouldn't stand out.

Accenting Results with Language

There are many ways to accent key words or results with language:

Italics

Quotation marks

One sentence paragraph

Short sentence at the end of a long paragraph

Repetition of parallel phrase or sentence structure

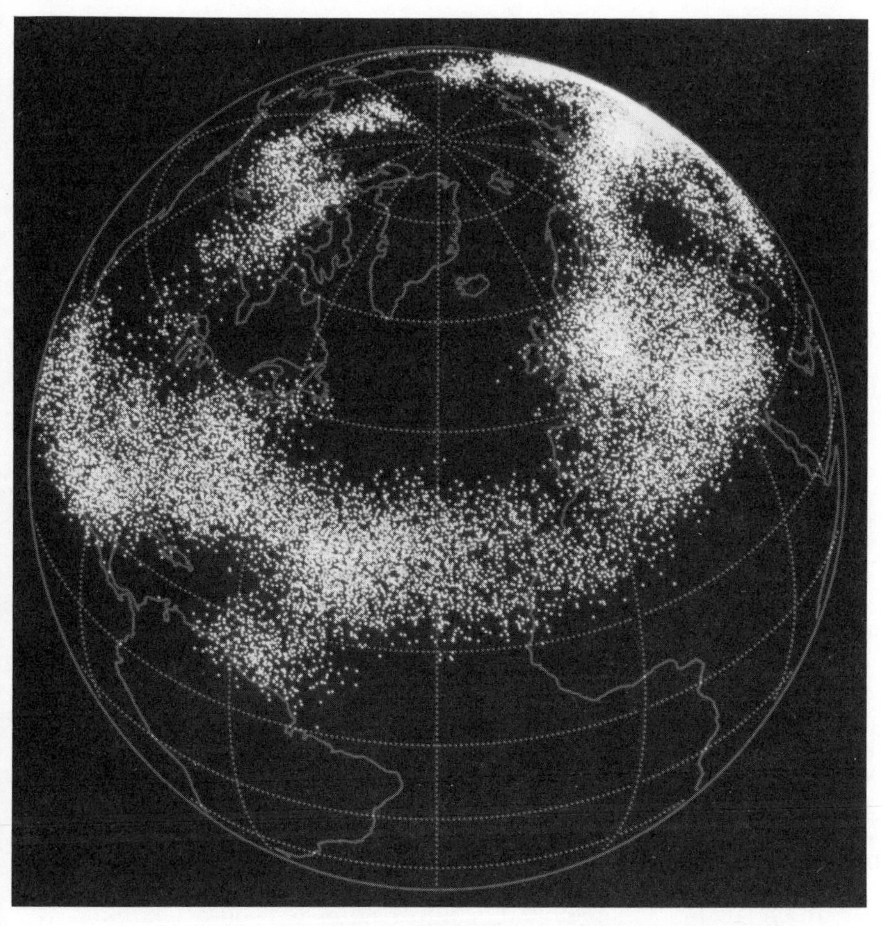

Figure 16-6. Computer-generated plot of smoke spreading around the globe five days after a nuclear exchange. Each dot represents about 4,000 metric tons of smoke spread throughout a volume of 550,000 km^3. The smoke is injected at five general areas in the U.S., Europe, and Soviet Union where large numbers of fires seem plausible.[7]

Use anything that calls attention to itself. The language device you choose is not that important. What is important is that you select something that makes your results stand out.

A qualitative picture of the longitudinal electric field at the axis of the electron avalanche is given in Figure 16-7. The x = 0 plane corresponds to the front of the avalanche and is

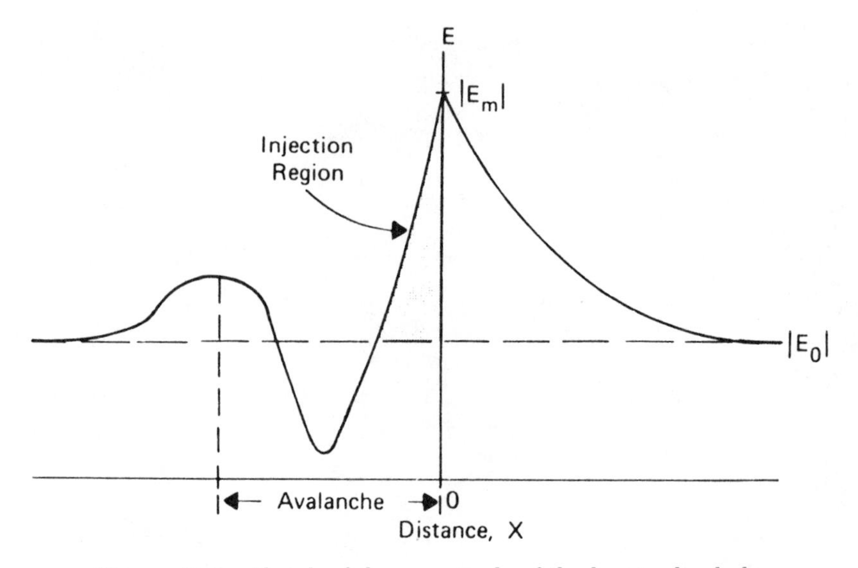

Figure 16-7. Sketch of the magnitude of the longitudinal electric field along the axis of the avalanche.

where the field attains the maximum average value it can have in the gap—E_m. Away from this plane, the electric field decreases. Notice that for $x > 0$*, the field decrease is slower than for* $x < 0$. **(illustration)**

The electron energy distribution is far from equilibrium. It is enriched with high-energy electrons, a consequence of the large electric field. Furthermore, the distribution is anisotropic in the high-energy region; it acquires a directed character along the field. We may think of the electron distribution as being formed by two groups of electrons: the "main" electron distribution and the "fast" electrons. **(quotation marks)**

Some of these fast electrons become runaways—that is, they continuously gain energy from the field. This gain occurs because the effective retarding force on an electron moving through a neutral gas decreases with increasing velocity (to run away, electrons need an energy of about $\simeq 4\epsilon_i$*, where* ϵ_i *is the ionization energy). The magnitude of the electric field determines the energy threshold for the electron runaway. The larger the field, the lower the threshold energy and consequently the larger the number of electrons which can runaway.* **(parallel phrases)**

Figure 16-7 also shows the "injection region," the high

electric field region where electrons are most likely to run away. In this "injection region," the runaway threshold energy is the lowest and is determined by the maximum field intensity E_m*. Because the distribution is anisotropic in the energy regime, the runaway electrons accelerate out of this injection region and inject into the region ahead of the avalanche where the field is decreasing.* **(illustration, quotation marks)**

The number of particles injected is maximum at the axis. Here, the space-charge fields and the external field are collinear; thus, the total field is at a maximum and the energy needed for injection is at a minimum. As we move away from the axis, the space-charge field and the external field are no longer collinear, so that the total field is reduced. Thus, the threshold energy needed for injection rises. We now observe that there is an injection cone with a maximum on the axis (see Figure 16-8). Once injected, most fast electrons no longer meet the runaway conditions and become "trapped"; that is, the energy they gain along their trajectory is not large enough to overcome their losses. For a given gas, the trapping distance (the distance from the avalanche to where the fast electrons become trapped) is a function of the initial injection energy of the electrons and the slope of the electric field ahead of the avalanche. **(quotation marks, illustration)**

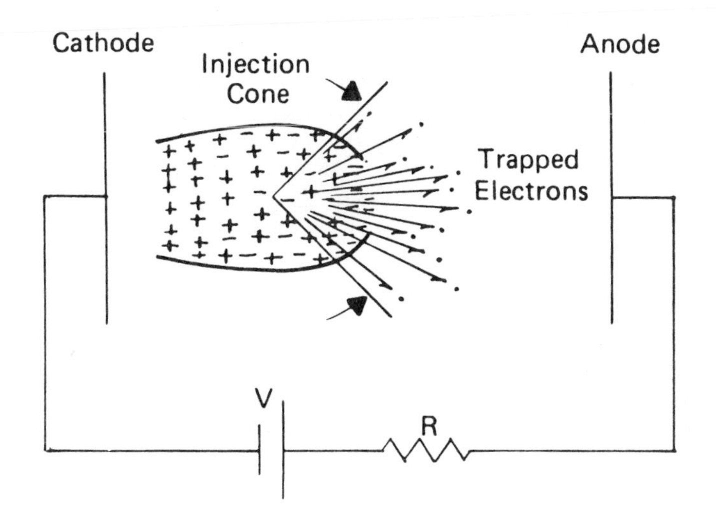

Figure 16-8. Injection cone at avalanche front. The electric field is strongest on the axis.

Therefore, just prior to Stage IV, we may think of the avalanche as a localized distribution of electrons. The beginning of Stage IV is marked by the "burst" of the avalanche along its axis, followed by the ejection of high-velocity electrons and their subsequent capture at varying distances from the original avalanche. At these points the captured electrons ionize the gas, thus extending the boundary of the main avalanche along a filimentary channel centered at the axis of the avalanche (see Figure 16-9). Once the channel starts to narrow, the electric field just ahead of the tip increases. Fast electrons are continuously injected into the region ahead of the tip and again are trapped. **(quotation marks, illustration)**

This process accelerates the development of the avalanche tip towards the anode. **(single sentence paragraph)**

Once the avalanche front makes contact with the anode, the field in the cathode side of the avalanche is greatly enhanced. This enhancement occurs because a conduction path now exists between the cathode tip of the avalanche and the anode. Moreover, the field lines converge toward this tip, so that subsequent avalanches emerging from the cathode can complete the final bridging of the gap. As the applied voltage increases, two things happen: (1) the avalanche "bursts" closer to the cathode; and (2) the number of untrapped runaway elec-

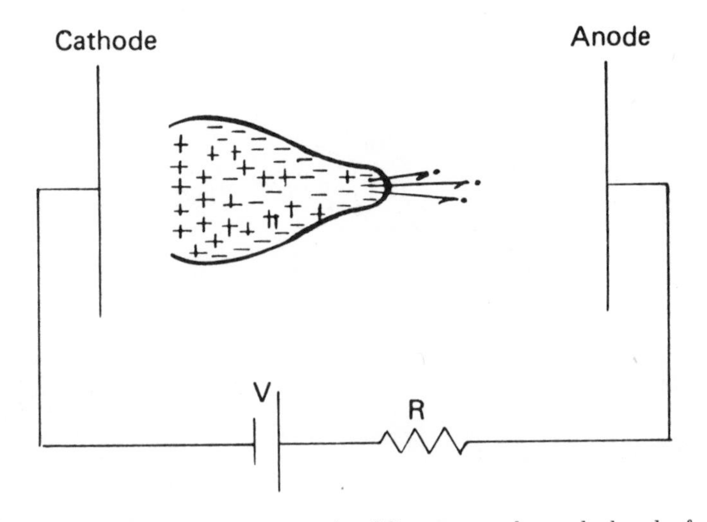

Figure 16-9. Development of a filimentary channel ahead of the avalanche.

trons increases.[8] **(quotation marks, parallel clauses, stylistic repetition)**

USING APPENDICES

Rarely will you write a report for only one type of audience. Most scientific reports have several different types of readers, each type with a different background and reason for reading the report. How then do you write your report for all these audiences? The answer is you don't. You write your report for your main audience, and you supply appendices for your secondary audiences.

Using appendices to give detailed information to secondary audiences. Let's assume you developed a computer code for modeling the chemical kinetics in a turbulent flame. Let's say your principal audience was a combustion research group who cared more about the way you described chemical reactions than about your computer code's iterative schemes. In your discussion, you would describe the physical assumptions of the computer model, but you would detail the iterative schemes of the program in an appendix. This appendix would be for the paper's secondary audience: computer programmers.

Using appendices to provide general background information to secondary audiences. Let's say your primary audience for a cotton pest management report was a technical audience (wildlife biologists) and your secondary audience was a nontechnical audience (cotton farmers). In the back of your report you might include a glossary for your nontechnical audience (as shown below). This glossary[9] allows nontechnical readers to understand the vocabulary of your report without breaking the fluidity of the writing for your technical readers.

GLOSSARY

Annual: a plant that normally completes its life cycle of germination, growth, reproduction, and death in a single year.

Anthers: the pollen-producing organs of flowers.

Axil: the angle between a branch or leaf and the stem from which it grows.

Blackarm: bacterial blight lesions on stems.

Canker: a dead, discolored, and often sunken area (lesion) on a root, stem, or branch.

Chlorosis: yellowing or bleaching of plant tissue that is normally green.

Crochets: tiny hooks on the prolegs of caterpillars.

Diapause: a period of physiologically controlled dormancy in insects.

Frass: a mixture of feces and food fragments produced by an insect in feeding.

Girdle: a ring of dead or damaged tissue around a stem or root.

REFERENCES

1. U. S. Department of Energy, *Solar Thermal Technology Annual Evaluation Report*, (Golden, CO: Solar Energy Research Institute, 1985), p. ii.
2. D. L. Siebers and T. M. Dyer, "Autoignition and Combustion of Coal-Water Slurry under Simulated Diesel Engine Conditions," Presented at the *Diesel and Gas Engines Symposium*, (Dallas, Texas: February 1985).
3. Pace VanDevender, "Ion-Beam Focusing: A Step Toward Fusion," *Sandia Technology*, 9, no. 4 (December 1985), pp. 2–13.
4. E. Kunhardt and W. Byszewski, "Development of Overvoltage Breakdown at High Pressure," *Physical Review A*, 21, no. 6 (1980), p. 2069–72.
5. Gary Carlson, "Implantable Insulin Delivery System," *Sandia Technology*, 6, no. 2 (June 1982), pp. 12–21.
6. M. C. Stoddard, *Convective Loss Measurements at the 10 MWe Solar Thermal Central Receiver Pilot Plant*, SAND85-8250, (Livermore, CA: Sandia National Laboratories, 1986), pp. 31–32.

7. Lawrence Livermore National Laboratory, "Global Atmospheric Effects of Nuclear War," *Energy and Technology Review*, (Livermore, CA: Lawrence Livermore National Laboratory, May 1985), p. 28.
8. E. Kunhardt and W. Byszewski, "Development of Overvoltage Breakdown at High Pressure," *Physical Review A*, 21, no. 6 (1980), p. 2069-72.
9. Western Regional Integrated Pest Management Project, *Integrated Pest Management for Cotton*, (Berkeley, CA: University of California, Division of Agriculture and Natural Resources, 1984), p. 142.

CHAPTER 17

The Ending

Science is built up with facts, as a house is with stones. But a collection of facts is no more a science than a heap of stones is a house.

J. H. Poincaré

The ending or conclusion section of your paper or report should repeat the most important results of your research, then discuss those results in the context of the whole research. You should not present any new results in a conclusion. Neither should you take a Perry Mason approach in which you bring in new details that unravel the mystery of your research. Instead, you should repeat your most important results and look at those results from an overall perspective.

How long should a conclusion be? For a short paper, a conclusion may be only one sentence.

> *These tests show that a nonpowered igniter for lean hydrogen-air mixtures is feasible, and that such an igniter could contribute to the safety of light water nuclear reactors by igniting safe concentrations of hydrogen during a loss-of-coolant accident.*[1]

Typically, though, conclusions run the length of an informative summary.

What's the difference between a conclusion and an informative summary? Sometimes, very little. However, a conclusion addresses an audience that has read the paper while a summary does not. A conclusion section also gives you a forum to make recommendations about future research:

> *These studies show that smoke from a nuclear war will reduce yearly global temperatures by 25°C for the Northern hemisphere. To confirm our initial assumptions about smoke injection, we need to measure smoke emissions from fires of various sizes.*

Conclusions bring together the loose ends of your research. Although you often cannot tie everything into a neat package in your conclusion, you can convey some sense of completion to your audience. In other words, you don't have to reach a summit in your conclusion, but you should arrive at a plateau. Consider a conclusion to a study for building an experimental solar power plant. The purpose of the experimental plant is to entice utilities to construct solar power plants on their own without government subsidy.[2]

> *In this study we found that six options for a solar power plant on a commercial scale (50 megawatts) were technically sound. These six options are described in Table 17-1. Although we found six technically sound options, we could not identify one option as a clear technical choice.*
>
> *Because we could not identify one particular option, electrical utilities should play a strong role in the final decision. The selection of an option will hinge on two choices by the utilities: (1) a stand-alone or hybrid plant; and (2) a molten salt or liquid sodium heat transfer fluid.*
>
> *(1) A hybrid plant with a fossil fuel source is an attractive option. Although a hybrid plant costs more, it assures plant operation even when the solar equipment is not operating.*
>
> *(2) The choice between liquid sodium and molten salt is difficult. Liquid sodium has about a 10% higher heat transfer capability than molten salt, and therefore should produce the lower costing receiver. However, molten salt has about a 25% higher storage capability than liquid sodium and should have the lower costing*

storage. The combination of a sodium receiver and molten salt storage is attractive, but is complicated by the heat transfer from one fluid to another. The cost of this heat transfer may negate any advantage of a two-fluid system.

Table 17-1
Solar Power Plant Options

Option	System Description
1	Salt Receiver and Storage Stand Alone
2	Sodium Receiver and Storage Stand Alone
3	Salt Receiver and Storage Hybrid
4	Sodium Receiver and Storage Hybrid
5	Sodium Receiver; Salt Storage Stand Alone
6	Sodium Receiver; Salt Storage Hybrid

If a successful plant at a power level of 50 megawatts will provide sufficient data to cause utilities to construct a plant without government subsidy, then any one of the options would become attractive. If the utilities indicate they need another experiment prior to their construction of a plant, then none of these options may be appropriate.

Notice that this conclusion not only repeats the most important result—that none of the options was a clear technical choice—but also considers that result from an overall perspective and makes a recommendation: that electric utilities play a strong role in selecting the option. Notice also that this conclusion contains an illustration (Table 17-1). Many scientists think that it's wrong to use an illustration in a conclusion (or in a summary, for that matter). You should use an illustration wherever an illustration is needed; and this particular conclusion needed a table to quickly identify the options.

REFERENCES

1. L. R. Thorne, J. V. Volponi, and W. J. McLean, "Platinum Catalytic Igniters for Lean Hydrogen-Air Mixtures," *Sandia Combustion Research Program Annual Report,* (Livermore, CA: Sandia National Laboratories, 1985), chap. 7, p. 12–14.
2. P. K. Falcone, *Handbook for Solar Central Receiver Design,* SAND86-8006, (Livermore, CA: Sandia National Laboratories, 1986).

PART FIVE

Actually Sitting Down to Write

CHAPTER 18

Getting in the Mood

A thermos of tea, a quiet room in the early morning hours. . .

Carson McCullers

No one can teach you how to write. Perhaps you can learn something by imitating the habits of other writers, but the actual act of sitting down to write is individual. What works for Meg Greenfield or George Will may not work for you.

Scientific writing is hard work. Granted, it's not as physically exhausting as swinging a pick or as mentally demanding as solving a nonlinear differential equation, but it requires concentration and patience. Moreover, the solutions are not exact. You don't write a draft of a paper and sit back and say, "Perfect." No matter how many times you draft a paper, there will always be some phrases that won't sound right, as well as some sentences where you feel compelled to state about ten details at once.

Scientific writing is lonely work. Although you may brainstorm ideas in a group or solicit critiques from other people, the actual process of sitting down to write demands solitary confinement. Just because scientific writing is difficult and lonely does not mean it is drudgery; not at all. Writing a strong scientific paper demands energy. You must convey complex images and ideas. Strong scientific writing also demands imagination. You must detach yourself from your research and place yourself in the position of your audience. Moreover, you will find that writing a strong scientific paper challenges not only your writing skills, but also your scientific skills. When you write to inform (not to impress), you see your own theories and experiments as your audience sees your theories and experiments. You become a critic of your own research. As Francis Bacon said, "Reading maketh a full man, conference a ready man, and writing an exact man."

How do you get started? You need two things before you can begin any scientific paper. First, you need something to write about; either some idea for a proposal, or some results from research. Although you don't have to wait until every result has been recorded, mapped, and plotted before you begin writing, you should not start writing a scientific paper that has no direction.

The second thing you need is an audience. You must have some idea about who will read your paper; what they know about your research, as well as why they are reading it. Without an audience, you can't efficiently map your trail; you don't know what background material to include, what words to define, how complex to make your illustrations.

Let's assume you've done some research and have an audience. Let's say you've spent the last fourteen months testing solid particles as a new heat transfer media for solar power plants. The simplicity of the idea intrigues you. You envision a solar central receiver plant in a desert; hundreds of mirrors surrounding a huge receiver; and silica sand falling through the receiver, collecting energy from the mirrors, and storing it in a tank below.

You've not only got visions; you've got results from simulation experiments in which concentrated sunlight heated silica sand particles to over 1000°C.[1] You've also finished a computer model that shows the concept's thermal performance to be about 75%,[2] a performance competitive with current central receiver designs. Your audience is varied: Department of Energy (DOE) officials, solar engineers, and utility engineers. Over the past few years, funding for the solar program has dropped; current solar central receiver designs have not proved efficient enough for utilities to invest in them. Management in the solar program is looking for new ideas, and you see solid particles as a good one. Unfortunately, some officials at DOE want a quick-fix solution; they're antagonistic to something long-term such as solid particles. Nonetheless, you see solid particles as the best chance that solar energy has for competing with fossil fuels.

Your supervisor sees something too because he is anxious for your report. "Have you written your report?" he asks you.

"No, not yet," you say.

Your supervisor stares at you for a moment. He has a casual look about him that's deceptive. His office suddenly feels hot. "I'll be seeing something soon," he says.

"Soon," you say. "Soon."

You go back to your office where Hank Wilson is waiting. Hank Wilson wants to talk about the Yankees. The Yankees are four games out of first place, and it is early September. You are not a Yankees fan—in fact, up until a year ago you hated the Yankees—but now Phil Niekro is with them and you'd like to see him pitch in a world series. Niekro is your hero. He is the only active player in major league baseball who is older than you.

Hank starts in about a catch Ken Griffey made against the Angels. "He never broke stride. You should have seen him. He dug his cleats into the wall and ran right up it. His glove must have been fifteen feet off the ground when he caught the ball."

You want to talk about baseball, but you can't. You've got results, damn it, and you've got to write a paper. What should you do? You can't write here with Hank talking about baseball. The library, you think. "Catch me on it later, Hank," you say. "I've got to write my report."

Hank looks disappointed. His look bothers you. If only

you didn't have to write, you think. You grab a pen and a pad of paper and start out the door, but then you stop. Should you use a pen or pencil? If you have to erase, you will need a pencil. On the other hand, you like the look of blue ink on yellow paper. You decide to bring both. You are almost out of the building when your secretary stops you. "Al put something in your mail slot," she says.

"I don't want to look," you say. Al is your department manager.

"Then don't," she says.

For one instant, you think about moving on. But curiosity gets the better of you. In your mail slot is the program's *Quarterly Report.* It's your turn to review it. The *Quarterly Report* is DOE's pride and joy. It contains almost nothing technical, only management things: milestone charts, procurement summaries, fiscal plots, and contract distributions. The kinds of things only DOE would want to read.

"By tomorrow," she says.

"Tomorrow?"

"Don't look at me. That's what Al said."

"But I've got to write my report."

"That's fine," she says. "As long as this gets reviewed by tomorrow." She smiles at you. Your secretary has a wicked smile.

You go back to your office. Hank Wilson stares bewildered as you walk by. You try to ignore him. In your office, you begin reading. You are only halfway through the standard DOE foreword when you catch yourself just staring at the words and sentences. You sit up in your chair, go back to the first sentence, and begin again. Thirty minutes into your reading, a utility engineer from Arizona calls. His company is studying solar energy options, and he wants to know the thermal efficiencies of solid particle receivers. You promise to send him something tomorrow. You begin reviewing the *Quarterly* again. Al expects you to do a good job. Competition for funds is tight. You push through the milestone charts and fiscal sheets and begin checking the procurement summaries when an engineer from analytical modeling stops by. She wants to adapt your computer code to solve a problem in coal combustion. You force a smile and invite her in. She has several questions about your code; questions you try to answer off the top of your head, but can't. So you dig through your old notes and printouts. By the time she leaves, it is five-thirty. You are tired and hungry. What now? You decide to go home and return here tonight, when no one is here and you can take the phone off the hook and write.

You get home and there's a bill from the power company. It's a bill you're sure you've already paid. You throw it in the waste basket; then you pick it out. There's something frightening about throwing away an unpaid bill. You decide to forget the bill and eat out. You need a good meal, you think. After all, you've got a lot of writing in front of you. You go to Pedro's for chili rellenos. While you eat, you read today's newspaper for the first time today. Guidry pitches tonight, Niekro tomorrow. Niekro's going for #298.

It's seven o'clock by the time you finish and get back home. You're a little tired and you decide to rest a bit; just lay down on the couch and think about what you're going to write.

When you wake up, it's dark and still outside. You look at the clock—11:43. The day is lost.

Scientific writing is a craft that requires preparation. It is very hard to jump right into writing a paper. Most professional writers have disciplined schedules. Although you as a scientist cannot alter your entire life style just to get one paper or report written, you can make some simple adjustments.

The first thing you should do before you begin writing is to clear your mind. You must forget about your personal problems, the bills you have due. You must think about your research and your audience. One good way to clear your mind is through motion: walking, running, mowing the yard, even driving. Wright Morris walks twice a day, before and after writing. Harper Lee plays golf. Max Apple and John Irving are avid runners. Mowing is okay, but you've only got so much lawn. Driving lets you daydream, but driving and daydreaming at the same time are dangerous. Running and walking are best; they give you a chance to think about the structure of your paper, to play a strategy through in your mind and see if it makes sense. I prefer a slow steady run, about thirty to forty-five minutes, not so long that I'm exhausted, but long enough to begin daydreaming. After taking a run or walk, I don't eat or sleep. I just shower and slip into something casual.

The second thing you need is a block of free time. For first drafts, a good writing block is three to four hours. Even after sitting down at your desk, you usually need twenty minutes or so just to get into the groove of actually writing sentences. Don't think though that you can go all day writing an early draft. First drafts are strenuous. You must juggle your audience and research against all three elements of style: structure, language, illustration.

Third, you need a quiet place; some place where the phone won't ring, where visitors won't sweep in and out, where there are few distractions. You will probably find that silence is not necessary—white noise is okay. Soft instrumental music is okay, too; but not television or hammering or children crying.

Finally, you must mentally prepare yourself for the task ahead. Scientific writing is hard work; it's not something you can casually start after a night on the town. Many scientists mistakenly think of professional writers as free spirits who write perhaps two days of the week and play the other five. Most professional writers are diligent workers. Hemingway's schedule, for example, called for eight hours of writing, then eight hours of leisure, then eight hours of sleep.

Instead of worrying about the way that professional writers spend their leisure time, you should concentrate on the way that professional writers spend their writing time. For instance, you should study how professional writers structure their sentences and paragraphs. Don't, however, choose writers from the eighteenth century whose language is outdated, or writers such as Faulkner or Joyce whose styles are often too luxurious for the first constraint of scientific writing. Choose a writer such as Hemingway whose style is crisp and clear.

They hanged Sam Cardinella at six o'clock in the morning in the corridor of the county jail. The corridor was high and narrow with tiers of cells on either side. All the cells were occupied. The men had been brought in for the hanging. Five men sentenced to be hanged were in the five top cells. Three of the men to be hanged were negroes. They were very frightened. One of the white men sat on his cot with his head in his hands. The other lay flat on his cot with a blanket wrapped around his head.

They came out onto the gallows through a door in the wall. There were seven of them including two priests. They were carrying Sam Cardinella. He had been like that since about four o'clock in the morning.

While they were strapping his legs together two guards held him up and two priests were whispering to him. "Be a man, my son," said one priest. When they came toward him with the cap to go over his head Sam Cardinella lost control of his sphincter muscle. The guards who had been holding him up both dropped him. They were both disgusted. "How about a chair, Will?" asked one of the guards. "Better get one," said a man in a derby hat.

When they all stepped back on the scaffolding back of the drop, which was very heavy, built of oak and steel and swung on ball bearings, Sam Cardinella was left sitting there strapped tight, the younger of the two priests kneeling beside the chair. The priest skipped back onto the scaffolding just before the drop fell.[3]

Although you don't want to copy everything about Hemingway's style, you should note how he relies on familiar language and simple sentence structures. There is nothing complex here—just short crisp sentences doing the job.

REFERENCES

1. J. H. Hruby and B. R. Steele, "Examination of Solid Particle Central Receiver: Radiant Heat Experiment," *Proceedings of ASME-ASES Solar Energy Conference,* (Knoxville, TN: March 1985).
2. C. A. LaJeunesse, *Thermal Performance and Design of a Solid Particle Cavity Receiver,* SAND85-8206 (Livermore, CA: Sandia National Laboratories, 1985), p. 56.
3. Ernest Hemingway, "Big Two-Hearted River," *The Short Stories of Ernest Hemingway,* (Philadelphia, PA.: Charles Scribner and Sons, 1976), p. 219.

CHAPTER 19

Writing the First Draft

A writer will be completely ruthless if he is a good one. He has a dream. It anguishes him so much he must get rid of it. He has no peace until then. Everything goes by the board: honor, pride, decency, security, happiness, all, to get the book written.[1]

William Faulkner

Tomorrow, you quickly take care of the small things at work: your critique of the *Quarterly Report,* a memo to the utility engineer in Arizona. You leave work early and eat a light meal at home. You wash the dishes, put in a load of laundry, and go out for a slow run; something to clear your mind. You run out beyond the subdivision, out into cattle land—barbed wire, brown grasses, gravel shoulder. It is still sunlight, and you are thinking about what you are going to write.

After you finish your run, you shower and drive back to the office. No one is there. It is almost seven o'clock. Somehow, the place seems different; maybe it's the stillness—or just the quiet. The only sound is the hum of the drinking fountains. You sit at your desk with your spiral notebook of results in front of you. You are in the mood.

You grab your pencil, put it down, then pick up your pen. You think about using the computer, but decide you need something written on paper first. You write a title for your paper:

Solid Particle Solar Central Receiver Design

Tiresome, you think. Too many big words in a row. You try again.

Solid Particle Receiver Design

That's too confusing; most people wouldn't know that you are talking about solar energy. More than that, the research isn't really the design of a solid particle receiver; it's deciding whether solid particles could work in the receiver. You write another title:

Using Solid Particle Receivers in Solar Central Receiver Systems

It's not great, but it will get you started. You write your name below the title. The rest of the page is blank. It stares back at you like snow on a television screen. Suddenly, you feel tired. You sit back in your chair and rub your eyes. You loathe first drafts. You would do almost anything before beginning a first draft: fill out your taxes, clean the bathtub, watch a stock car race on television.

"Come on," you tell yourself. You've wasted enough time. You must start. You must write something down; an outline, you think. That's what you were always taught. So you write one:

A. INTRODUCTION
B. EXPERIMENT
C. COMPUTATIONS
D. CONCLUSIONS

What about a summary? You write SUMMARY in above INTRODUCTION. What letter should you assign it? You already used A on INTRODUCTION. You decide to make it Z. You're not happy with Z but you see little choice besides crossing out the other four. You wish you had used a pencil. Now you look at your outline.

Z. SUMMARY

A. INTRODUCTION

B. EXPERIMENT

C. COMPUTATIONS

D. CONCLUSIONS

Not much help, you think.

You decide to try something else. You write down SUMMARY at the top of a blank page, then INTRODUCTION at the top of another page, then EXPERIMENT, COMPUTATIONS, and CONCLUSIONS at the top of three more. You pick up your five outline pages as well as your title page and bunch them in a stack, then place them back down. Now, you're getting somewhere.

You put aside the title page and start with the summary page, but you remember that summaries are the last thing you write. You put it aside and look at your introduction. You start writing a sentence about how the solar central receiver program began in 1976, but then you stop. What's an introduction supposed to do? Introductions are supposed to prepare readers for what's ahead. How can you prepare someone for something ahead if you don't know what that something is? You put the INTRODUCTION page aside also.

So far, things aren't going well.

For some reason you are restless. It is eight o'clock and you are hungry again. You go down the hall and get a candy bar and diet soda. You know candy bars are bad for you, but you're working hard tonight and you deserve a treat. After you finish the candy bar, you feel tired. You think about calling Hank Wilson to get the Yankees' score. Then you think about going home and crawling into bed and sleeping until you can't sleep any more.

You have lost the mood.

Don't deceive yourself. Early drafts are difficult; very difficult. You must juggle your research and audience with all three elements of style. What can you do? The first thing you can do is relax. You're not going to finish your paper in one sitting; and if you do, chances are it won't be any good. You've got to be realistic. It has taken you months to do the research. It's going to take some time to finish the report—maybe three weeks, maybe four.

Now, for the actual act of writing the first draft. Should you use a pencil or pen? As John Gardner said, that question is much deeper than it sounds.[2] The real question, he says, is how do you get words on paper. Regarding the surface question of what writing instrument you should use, the answer is whatever it takes. If you feel more comfortable with a pencil, use a pencil. If you feel more comfortable on a computer, then use a computer. If you feel more comfortable etching your words on stone tablets, then go to it.

What about that deeper question of how you get words on paper? The answer is the same: whatever it takes—short of plagiarism. Should you use an outline? Every English teacher you've ever had has probably told you to use one. That fact alone should make you suspicious. The answer to the question is a qualified "yes." A strong outline is essential for complicated scientific papers. A strong outline gives you a structure before you begin writing. That way, in your early drafts, you have to juggle only two elements of style—language and illustration—not three. To be effective, though, your outline must be strong. What makes for a strong outline? The answer is detail. A strong outline not only lists your section headings, but also your subsection headings, notes about each section, and any sentences or phrases you might use. Strong outlines look nothing like the neat little ABCD things your teachers always put up on the blackboard. Strong outlines (such as the one shown) are long and spread out and, above all, are filled with the kinds of things that make it easy for you to write a first draft.

OUTLINE

USING SOLID PARTICLE RECEIVERS IN SOLAR CENTRAL RECEIVER SYSTEMS

Audience

Principal audience is DOE headquarters who has general knowledge about solar central receivers, but not about solid particle research; some DOE members are antagonistic to solid particle idea (they want quicker solution).

Secondary audience includes solar engineers at national laboratories and utilities who have fairly detailed knowledge about solar central receivers and mainly want to know if the solid particle idea can work.

INTRODUCTION

Identity of research:

This report will discuss using solid particles as the heat transfer medium in a solar central receiver . . . . Requirements for such a heat transfer medium include the following:

(1) *Temperature of heat transfer medium must reach 800°C.*
(2) *Convective losses must not be too high (define "high")*
(3) *What else???*

Reasons why research is important:

(1) *Diversifies solar energy applications by obtaining higher temperatures than those obtainable from current solar designs.*
(2) *Allows for high-temperature experiments and modeling.*
(3) *Maybe go into importance of central receivers?*

Background material:

(1) Water/steam. *The advantages: utilities familiar with; test data available. The disadvantages: not efficient enough; difficult to control in two-phase flow.*
(2) Molten Salt. *The advantages; relatively high heat capacity; high storage capacity; single-phase fluid. The disadvantages: freezing in pipes; corrosive nature.*
(3) Liquid Sodium. *The advantage: high heat transfer capacity. The disadvantages: low storage capacity; safety problem.*

(4) Solid Particles.
Advantages:
-Very high heat capacity (define "very high")
-Can absorb irradiation directly
-Can transfer heat directly
Disadvantages:
Abrasive nature
Utilities less familiar with transporting

DISCUSSION

.
.
.

How do you actually write the first draft? Let's first assume that you have a strong outline and that you are in the mood. Now, there are two approaches: the rabbit's and the turtle's.

Rabbits hate first drafts. They hate juggling audience and research with all the elements of style. In a first draft, they sprint; they write down everything and anything. They use pens—no time for erasing—and their pens never leave the paper. Rabbits strap themselves to the chair and will not get up for anything. Rabbits finish drafts quickly, but their early drafts are horrendous, many times not much better than their outlines. Nonetheless, they've got something. They've got their ideas on paper, and they're in a position to revise.

Turtles, on the other hand, are patient. Turtles accept the job before them and proceed methodically. A turtle tries not to write down a sentence unless it's perfect. In the first sitting, a turtle begins with one sentence and slowly builds on that sentence with another, then another. In the second sitting, a turtle then goes back to the beginning and revises everything from the first sitting before adding on. It usually takes a turtle several sittings to finish a first draft, but the first draft is strong. When a turtle finishes a first draft, the beginning and middle are usually very tight because they've been reworked so many times. Revision usually entails smoothing the ending as well as checking the overall structure.

Which type of writer should you be? The answer is whatever type works for you. Actually, no writer is strictly rabbit or strictly turtle. Most writers are a combination. If your writing slows to a stop, you must become a rabbit or you'll never finish. If you're working on a section that establishes the structure for the rest of your paper, you should be a turtle; otherwise, you might have to start over.

On which section of a paper should you start writing? The introduction or the discussion? This question is tough. Psychologically, you need some kind of introduction before you can begin the discussion. But the discussion is the easiest section to write first. My feeling is that you should be a rabbit in your introduction, then write the discussion and ending at your own pace, and then go back and rewrite your introduction. What about your summary? If your summary is descriptive, you may want to write it first; but if your summary is informative, you definitely want to write it last. One good way to tackle the informative summary is to wait until you have a polished draft of your report, then highlight its most important sentences and illustrations, and then extract those sentences and illustrations for the first draft of your summary.

The key to writing first drafts is momentum. Because scientific writing is so *textured* (because each sentence depends so much on the ones around it) you need to sustain momentum during your first draft. Otherwise, you'll lose your train of thought. How do you sustain momentum? Well, there are a few habits you can borrow from professional writers.

First, always write in blocks. Set a goal for yourself at each sitting. Take on only one or two sections. How large a goal you set depends upon how much of a rabbit or turtle you are. Realistic goals are the best. Then, once you set a realistic goal, achieve it even if you have to finish the section in outline form. Psychologically, you want to end each sitting with a feeling of accomplishment.

Second, never end a sitting with the last period of a section unless it's the last period of the paper. Before finishing a writing session, always begin writing the next section. That way it's easier to start writing the next time you sit down. The fiction writer André Dubus claims he always stops mid-sentence, mid-thought.

Third, watch what you eat. Writing makes you restless, which makes you hungry. Don't eat foods that will make you sleepy or will stain your papers. Also avoid salty foods that make you thirsty and foods that require both hands such as bananas (you have to peel them). You want something that you can reach over and grab with your nonwriting hand. Celery and carrot sticks are probably the best, but you can munch on them for only so long. Raisins aren't too bad; neither are unsalted sunflower seeds. I sometimes have a trail mix: raisins, unsalted peanuts, shredded coconut, granola, and carob.

Finally, when you finish an early draft, get it typed and store away a clean copy. Having a clean version that you can pull out any time gives you psychological advantage when you start revising the original. You won't worry about being caught empty-handed in case your supervisor needs a clean draft right away.

How do you write a first draft when you're pressed for time? When pressed for time, the one part of the writing process you don't want to cut back on is your first draft. If anything, you should spend extra time on your first draft because your first draft determines your structure. When you're in a hurry, you need to select a structure that works on the first go-round. Otherwise, you could have your paper or report in pieces on your desk when the deadline strikes. Therefore, in your first draft, be conservative. Spend time developing a structure that you're sure your audience will follow.

There are really no shortcuts in a first draft outside of abandoning your family and locking yourself in a room until it's done. (That practice, however, usually causes more grief than anything else.)

How do you write a first draft when you have several people collaborating? Multiple authorship poses some potential problems: illogical strategy, weak transitions between sections, inconsistencies in language. To avoid these problems, you should designate someone—preferably your best writer—as coordinator. This coordinator assumes responsibility for the outline, introduction, summary, and conclusions. This coordinator also has the license to change any section in order to smooth its transition into the whole. The coordinator is not a dictator. All contributors should have the chance to comment on at least one revision.

What do you do if you've got *writer's block*? Writer's block is one of those bits of folklore that scientists love to talk about, especially when they're having problems writing. "The words just aren't flowing," they'll say, "I must have writer's block." Well, the words rarely "flow" for anyone (professional writers included). Most professional writers struggle the same way you do. They struggle with paragraphs, sentences, and words. When writing a scientific paper, there are three things, besides distractions, that cause you to stop:

1. *You can't think of the right word.* Not having a word happens to everyone. Many times it's on the tip of your tongue. A possible solution is to pull out a dictionary and spend five minutes looking. If you don't find the word after five minutes, write down the name of your favorite baseball team and keep going. Your subconscious will think of the right word later.

2. *You can't find the sentence to express an idea.* Chances are you haven't actually grasped the idea yet. A possible solution is to skip two lines and keep writing. Let the idea simmer. Maybe you can think about it on your next run or walk. Running and walking are excellent ways to sharpen ideas.

3. *You hear voices.* When many scientists sit down to write, they hear voices, critical voices: their eighth grade English teacher, their department manager, the theorist from Oxford who complains bitterly about everyone else's writing but doesn't write so well himself. These voices inhibit you—you just can't make your pen write another word. A possible solution is to put on your favorite classical or jazz record and turn up the volume until it drowns the voices. Two suggestions: Copland's *Appalachian Spring* and Jean Michel Jarre's *Oxygene.*

In scientific writing there really is no such thing as writer's block. Once you have done the research, the ideas are there. Writing those ideas is more mechanical than inspirational. For scientists, the real block occurs when the ideas won't come. But that block is a block in science, not in writing.

REFERENCES

1. Excerpt from interview with William Faulkner, from *Writers at Work: The Paris Review Interviews*, First Series, edited and with an Introduction by Malcolm Cowley. Copyright © The Paris Review, Inc. 1957, 1958. Reprinted by permission of Viking Penguin, Inc.
2. John Gardner, *On Becoming a Novelist*, (New York: Harper & Row, Publishers, Inc., 1983), p. 119.

CHAPTER 20

Revising, Revising, Revising

Fifteen years ago in Spain, I used to greet each morning spitting blood in the washbasin, having the night before gnashed the inside of my mouth while dreaming I had misplaced a comma in my writings of that day, throwing off the pattern of speech given to a character who lived two hundred years ago.[1]

Thomas Sanchez

Revision is the key to strong scientific writing. Revision is the difference between a title such as

> *Using Solid Particle Receivers in Solar Central Receiver Systems*

and a title such as

> *Using Solid Particles as the Heat Transfer Media in Solar Central Receivers*

For this research, the second title is more precise. You weren't deciding whether solid particle receivers could work in a solar central receiver system; you were deciding whether solid particles could work in a solar central receiver. This is an important difference; but this difference may not have been apparent when you were staring at the blank page of a first draft.

In your first draft, you have to juggle too many elements of style to give enough attention to any one element. Revision gives you the luxury of considering on any single draft only the clarity of your illustrations or the conciseness of your language or the parallel structure of your discussion. Scientists, perhaps more than other professionals, understand the principles behind revision. In their experiments, scientists depend on trial and error. In their theories, scientists use many iterative techniques to arrive at solutions. Just as those techniques are important to strong experiments and theories, revision is important to strong writing.

Many scientists hold the misconception that great writers don't have to revise. These scientists assume that great writers think only great thoughts and then effortlessly write them down in prose that needs no reworking. Well, maybe there are great writers such as D. H. Lawrence whose first drafts can go straight to press; but Lawrence is not in the majority. For every D. H. Lawrence, you can find ten writers who were constant revisers: Ralph Ellison, Leo Tolstoy, Flannery O'Connor, John Gardner, Carolyn Chute, Evan Connell, Raymond Carver, E. B. White, William Gass, William Kittredge.

Raymond Carver sometimes takes as many as thirty drafts to finish a story; never less than ten or twelve. If great writers such as Carver need ten drafts to smooth their writing, just think how many more revisions the rest of us need. In my writing, I average about ten pages a day. Unfortunately, they're all the same page. It's rare for me to have a sentence that was written on the first draft make it to the final draft unchanged. The key to successful writing lies in working hard on your revisions, not in conjuring any magic on your first drafts.

What does revision entail? Before you begin revising a paper, you should obtain some distance from it. After finishing a first draft, you should spend a day hiking in the mountains or a night on the town. Watch a movie or go see a ballgame. You need a few hours (sometimes a few days) away from your first draft before you can effectively revise. Things have to settle. During this time, ideas for changes will probably come to you. Write these ideas down, but don't go back to your draft.

Okay. Let's assume you have some distance between you and your draft. Now what? The first thing you should do is change your personality. You cannot have the same personality when you revise as you did when you were writing the first draft. It turns out that you do need something of a split personality to be an effective writer. Whether you worked as a rabbit or turtle on your first draft, you were always looking at your draft with the same question: *How can I build a trail to lead my readers through my research?* In the revision stage, you look at your draft with a different question: *How can I make my trail more efficient?* In your first draft, you build. In your revisions, you polish.

To be a successful reviser, you need to be a strong reader.

How do you become a strong reader? One way is to listen to other strong readers. When you've finished a clean draft of a paper, find two or three readers whom you respect—don't ask any "yes men"—give them a copy, and tell them who your audience is. Then listen. Just as no two people write the same way, no two people read the same way. Some people are particularly attuned to language; others are sensitive to overall structure. Listening to strong readers review a paper will help you strengthen your own reading.

A second way to become a strong reader is to practice. Work hard on your critiques of other people's papers. Be specific. Don't just tell someone a section is unclear. Show them why you got confused. Don't just mark the things you find weak in a paper; also make notes of the things you find successful. You will find that telling someone what is right with a paper is more difficult than pointing out the things that are wrong. In a scientific paper, you can always find small weaknesses, places where the writing could have been tighter or more precise, but you can't always articulate what makes a paper succeed.

How quickly you mature as a writer will depend on how quickly you mature as a reader.

Now, for the actual act of revising. What's the best method? There is no definitive answer to this question. There are, however, some techniques that many successful writers adopt.

First, when you start revising your own work, don't do it in the same place where you wrote the original draft. A new setting will help you obtain some distance from your draft. Also, if at all possible, have your paper or report typed on a word processor; that way, you won't be reluctant to change things. Avoid revising directly on the processor; the in-text commands make reading difficult. Revise instead on a clean, double-spaced copy, then incorporate the changes into the file later. Seeing a printed version of your paper makes the reading fresh.

Second, try to work through the entire draft in one sitting. Revising all the way through in one sitting makes your writing smooth. You'll see gaps in logic and faults in overall structure. During your first few revisions, you'll need long blocks of time because you'll be writing a lot—filling in spots you sped through in your first draft. If your report is long, you'll probably need more than one sitting for early revisions. Use the same system as you did with your first draft: block out a section or two, then work through that block without stopping.

Third, get some distance between each revision. Don't make two revisions in the same afternoon. You'll go stir crazy and start changing things for no reason. How many times should you revise a paper? For your paper to really sing, you should revise until you are nitpicking over language: a word here, a comma there. When you think your paper is about right, read it *aloud*. You'll be surprised at how many little things you'll catch with your ear.

Finally, solicit criticism for your writing. Soliciting strong criticism is an excellent way to revise. No matter how hard you try to separate yourself from your research, you will take things for granted. For instance, in your paper about solid particles, maybe you forgot to mention the composition of your solid particles in the summary (your solid particles were silica sand). That detail would be an easy oversight for someone engrossed in the research, but one that a fresh reader would catch right away.

One caution about soliciting criticism: Don't solicit criticism if you're unwilling to make changes. This caution may seem silly, but you wouldn't believe how many times someone has asked me to review a paper and then been appalled when I suggested changes. One time, a computer scientist went so far as to throw a tantrum over a simple usage ruling—he had used the archaic "an historical" instead of "a historical." This scientist wanted no criticism, only praise.

Once you solicit criticism of your writing, you must prepare yourself to accept that criticism. Don't be defensive when people whom you respect are harsh about your writing. They're not attacking you. Actually, when people become angry over your writing, it's a compliment of sorts. It means they're frustrated—they're genuinely interested in your research and want your writing to inform. If your critics are indifferent, that's the occasion to be upset. What do you do if your critics tell you the paper is good? First, be skeptical. Ask them to be specific. You're not fishing for compliments here; you just want to hear which results came across. If they don't mention an important result, chances are you need to accent that result more.

How do you know when to incorporate a criticism and when to dismiss it? This question is difficult. You shouldn't give a rubber stamp to every suggested change, even if the suggestion comes from a good reader. Weigh all criticisms. If you think a criticism is valid, incorporate it. If you think a criticism is off the wall, at least mull it over. Maybe there was a deeper problem with the writing that your critic just couldn't articulate.

Another caution about soliciting critiques: Don't seek too many critics. You'll get confused. On a five-year plan for one of its solar programs, DOE decided that everyone in the program should review the document. In effect, DOE wanted all the readers to become the writers. The result was chaos. Because the report was general to start with, the two hundred critiques only served to push it in two hundred different directions. For several years, DOE haggled over draft after draft of that report, wasting thousands of man-hours, before finally scrapping it.

For most papers and reports, two or three strong reviewers should suffice. Balance your reviewers, though. For a paper with a wide audience, don't choose three experts on your research mountain. Instead, choose readers with a variety of backgrounds: one scientist on your research mountain, one scientist from another research mountain, and (if possible) a technical writer. Don't assume that once a professional writer passes his or her hand over your writing, your writing is perfect. During one reading, a professional writer can suggest only so many changes. Before a professional writer can really polish your writing, you should already have your writing smooth. Otherwise, the professional writer must edit with a power saw instead of sandpaper.

How should you handle revision on a paper or report that has multiple authors? This question is also difficult. On one hand, you don't want to slight any contributors; but on the other hand, you've got to realize that no two people write the same way. As was stated earlier, someone should coordinate the project, someone with a license to change things as he or she sees fit. In the revision process—preferably early—this coordinator should allow everyone to comment on the whole paper. However, there should be only one hand (the coordinator's) in the final editing.

What happens if you are on a tight schedule? In a scientific research project, it's difficult to schedule enough time for writing. Research is unpredictable. Let's say you make a schedule in January:

1. six months to perform the experiment,
2. two months to analyze the results, and
3. one and a half months to write the report.

Well, that schedule sounds fine until you lose a week in February because of a vacuum leak, two weeks in April because of a late shipment, a week in May because of a sick technician, then another week in July because the computers were down.

The month and a half you scheduled for writing has slipped to one week. Okay, so what do you do? Well, as was stated in the last chapter, you should spend enough time on your first draft to secure a strong structure. Otherwise, your report could be in pieces on your desk when the deadline strikes. When you're pressed for time, you've got to revise efficiently. Revising a paper is much like leveling a piece of land. In your first couple of revisions, you wield a pick for groundbreaking as you edit your structure. Then, in your last few revisions, as you edit your language and illustrations, you use a rake for smoothing. When time is short, you can't afford to use a pick in your last couple of revisions. If you do, you might improve the overall grade a little, but you'll unsettle the soil a lot. Also, when pressed for time, you need to condense the time between revisions. This condensing does not mean that you cut out playtime between drafts; it means that in the time you have you play harder. To revise effectively, you must separate yourself from previous drafts. Otherwise, you'll go stir-crazy and start changing things for no reason.

At some point in the writing process, you have to say "enough"—enough to outside suggestions, enough to revisions. You have to decide that your paper is successful, and then you have to finish it.

Many scientists think that you can achieve perfection in writing. Except, perhaps, for the ten commandments, there is no perfection in writing—any kind of writing. Words are not exact substitutes for thoughts. No matter how many drafts of a paper you write, you will never achieve perfection. There will always be something you'll want to change: some sentence that won't sound right on Tuesday, another sentence that won't sound right on Wednesday.

Although there is no perfection in writing, there is success. For your sanity, you must find a point at which to stop drafting your papers. How high you define success will determine how much you polish your writing. Evan Connell once said he knew he was finished with a short story when he found himself going through it and taking out commas, then going through it again and putting commas back in the same places. Strong scientific research deserves that amount of care.

Although you can't achieve perfection in your writing, you should still strive for it. Strive for perfection, but be content with success. One of the beauties of writing is that you never stop learning. With each paper, you improve your craft.

REFERENCES

1. Tom Jenks, "How Writers Live Today," *Esquire*, 104, (August 1985), p. 124.

CHAPTER 21

Finishing

It ain't over till the fat lady sings.

Dick Motta

It is two o'clock in the morning. Your report has gone through the review board and you have finished proofing it for what seems like the hundredth time. You are exhausted, but you correct the last two typos and print out the final copy. No one is in the office. You wish someone was; anyone. He wouldn't have to read your paper, just look at it, admire how clean it is. You'd tell him the writing was smooth, read him the title and summary.

Finishing a paper or report is taxing work. By your final draft, the words seem dead on the page. You don't really read them; you only see yourself reading them. Although the final proof of a manuscript is often tedious, you can't let up. A small mistake such as a misspelled word will unsettle your readers and undercut the authority of your work. Finishing a paper is much like finishing a baseball game. Some teams, when they're ahead, let up during the last few innings. They play sloppily, sometimes so sloppily that they lose their lead. Some writers are the same way: they work hard on the first four drafts, then let up on the final draft and let typos pull down their work.

Writing a paper is thankless work. By the time your paper

appears in print, any feeling you have for the writing is usually lost. You are probably thinking about your next research project. That's the way the writing/publishing game is. Don't expect satisfaction in your writing to come from other people. You'll be sorely disappointed. Satisfaction in your writing has to come from within; it is the realization that you've done good work.

You make a copy of your report and leave it on your supervisor's desk for final sign-off. You are euphoric and exhausted at the same time. It is over. Your supervisor may require some small changes, but as far as you're concerned you've done your best and it's behind you. You leave the lab. The night is cool, and the moon full behind some thin clouds. You think you hear opera. You get in your car—a maroon and white Monte Carlo with a bad muffler—and drive past the laboratories, past the open fields tinted purple and blue by the night. At a stop sign, your muffler wakes a neighborhood dog. You wish you could wake everyone and tell them what you've done.

You're tired, but you can't go to bed; not just yet. You need some kind of celebration; something, anything. The Donut Wheel, you think. You'll get an old-fashioned plain and some coffee and read the sports pages. Niekro pitches again tomorrow. This time he's going for #300 and you've got a good feeling he's going to win.

APPENDIX

Ten Common Grammar and Punctuation Mistakes

1. Subject-verb disagreement

When using a singular subject, use a singular verb.

> *The* belt *surrounding the earth's Pacific plate* **accounts** *for 80% of all earthquakes and volcanoes.*
> *A* series *of shocks often* **precedes** *a large earthquake. (One "series.")*

If a subject consists of two or more singular nouns joined by *or, either . . . or,* or *neither . . . nor,* the subject is singular and requires a singular verb.

> *Neither* oxygen *nor* nitrogen **is** *a noble gas.*

When using a plural subject, use a plural verb.

Most earthquakes **occur** *along narrow belts separating the large plates of the earth's surface.*

The data *on precursor activity of earthquakes* **are** *difficult to analyze. (The word "data" is plural, as are other words such as "*phenomena*" and "*strata*.")*

Three series *of reports* **are** *going to be published. (More than one "*series*.")*

If a subject consists of two or more nouns joined by *and*, the subject is plural and requires a plural verb.

Germanium *and* silicon **are** *semiconductors.*

If a subject consists of two or more plural nouns joined by *or*, *either . . . or*, or *neither . . . nor*, the subject is plural and requires a plural verb.

Neither ceramics *nor* gases **conduct** *electricity at low voltages.*

2. Comma Splice

A comma splice is a comma that joins two independent clauses. You should replace comma splices with either periods or semicolons.

Incorrect:

There is no cure for Alzheimer's disease, it brings dementia and slow death to thousands of Americans every year.

Correct:

There is no cure for Alzheimer's disease; it brings dementia and slow death to thousands of Americans every year.

Comma splices often occur before connectives such as "however," "moreover," and "therefore."

3. Ambiguous pronoun reference

A pronoun refers to the last noun used. You can stretch this rule somewhat, but not to the point of introducing ambiguities into your writing.

Fires caused most of the damages in the three earthquakes. They lasted for several days and caused many deaths.

What does the "they" refer to? The "earthquakes"? The "damages"? The "fires"? Granted, most readers can figure out that the pronoun refers to "fires," but only after first being tripped.

4. Dangling modifiers

When a sentence begins with a participle phrase, you must make sure the phrase modifies the subject of the sentence. Otherwise, the phrase dangles.

Neglecting costs, Option A is the best choice because of its high thermal efficiency.

Did Option A "neglect" the costs? No, but the placement of the participle phrase "neglecting costs" makes the sentence read that way. You should group introductory participle phrases with the nouns they modify.

Neglecting costs, we found that Option A was the best choice because of its high thermal efficiency.

5. Dropping the last comma in a series of three or more terms.

The established rule for punctuating commas in a series is as follows: *in a series of three or more terms with a single conjunction, use a comma to separate each term.* Therefore, write

carbon, silicon, germanium, and selenium
neopentane, perdeuteroneopentane, or neooctane

Poets and fiction writers sometimes leave off the last comma in a series of items. The reason? Rhythm—leaving off a comma slightly alters the pauses in a sentence. In literature, rhythm is sometimes as important as clarity. In scientific writing, however, clarity is always more important than rhythm. Any half-pause you may save omitting the last comma in a series is lost in a complex word such as "perdeuteroneopentane." Moreover, leaving off the last comma in a series often causes ambiguities.

The system consists of a black cavity radiometer, flexible cable to move the instrument through the flux field and associated data acquisition electronics.

This sentence trips the reader. Did the system contain the associated electronics, or did the instrument move through the associated electronics? A comma is needed after "field."

The fourth chapter gives major milestones, management organization and resource and procurement summaries.

Does the fourth chapter include management resource or resource summaries? A comma is needed after "organization." In both of these examples, the reader could have stopped and figured out the correct punctuation. The point though is that the reader has to stop.

6. "Which" for "that," and vice versa

Use "that" for defining clauses. Use "which" for nondefining or parenthetical clauses.

We will select the option that has the highest thermal efficiency. (Tells which one.)

We will select Option A, which has the highest thermal efficiency. (Adds a fact about the known option.)

If you can omit the clause without crippling the sentence, use "which." Otherwise, use "that." Note that "which" clauses require enclosing commas. "That" clauses do not.

7. Misplacing quotation marks

Closing quotation marks should go outside of periods and commas.

Einstein said, "God does not cast the die."

J. A. Dirks, "Southwest Utility Expansion Plans: Implications for Solar Thermal Electric Technologies," Sandia National Laboratories, Livermore, CA, SAND85-8022 (1986).

8. Misforming possessives

Form the possessive singular of nouns by adding 's, no matter what the final letter is.

Weiss's theory
the octopus's eyes
Rayleigh's formula

The possessives of the pronouns *hers, theirs, yours, ours,* and *its* have no apostrophe. Other exceptions include a few proper names such as *Jesus* and *Mount St. Helens.*

Jesus' parables
Mount St. Helens' eruption

9. Hyphenating compound words

Compound words are common in scientific writing. Sometimes you can find the correct spelling of these compounds in the dictionary. Many times, though, you can't. In such cases, you must decide whether to hyphenate. Should you write *fly ash* or *fly-ash? flow field* or *flow-field? cross section* or *cross-section?* There are no clear-cut rules here. Many compounds start out as two words, then acquire hyphens after years of use. Although no clear-cut rules exist, two principles should guide you when confronted with a new compound.

The trend in spelling compound *nouns* is away from the use of hyphens. Hyphens make the writing seem more complex. Therefore, write

cross section
flow field
fly ash

When compounds appear as *adjectives* before nouns, use a hyphen to avoid misleading the reader. Therefore, write

cross-section measurements
flow-field predictions
fly-ash modeling

10. Capitalizing every part of an experiment, theory, or design

Capital letters also increase the complexity of writing. Although you must use capital letters for proper nouns and the beginning words of sentences, you should use lower case letters whenever possible. Don't write

> *In our Liquid Sodium Receiver Experiment, a Power Production Phase will follow the Test and Evaluation Phase.*

This sentence intimidates readers. Instead, write

> *In our liquid sodium receiver experiment, a power production phase will follow the test and evaluation phase.*

This revision is much easier to digest.

Bibliography

Grammar and Punctuation References

A Manual of Style, 12th ed. pp. 103–430, Chicago: The University of Chicago Press, 1969.

Theodore M. Bernstein, *The Careful Writer.* New York: Athenum, 1967.

Gorrell, Robert, and Charlton Laird, *Modern English Handbook* (6th ed.), Englewood Cliffs, N.J.: Prentice-Hall, Inc., 1976.

William A. Sabin, *The Gregg Reference Manual* (5th ed.), New York: McGraw-Hill Co., 1977.

Strunk, William, and E. B. White, *Elements of Style,* New York: MacMillan, 1979.

Format References

Council of Biology Editors Style Manual (3rd ed.), American Institute of Biological Sciences (3900 Wisconsin Avenue NW, Washington, D.C. 20036).

Handbook for Authors of Papers in the Journals of the American Chemical Society. American Chemical Society Publications (1155 Sixteenth Street NW, Washington, D.C. 20036).

Information for IEEE Authors. The Institute of Electrical and Electronics Engineers (Editorial Department, 345 East Forty-seventh Street, New York 10017).

Style Manual for Guidance in the Preparation of Papers for the Journals Published by the American Institute of Physics and Its Member Societies (rev. ed.), 1970. American Institute of Physics (335 East Forty-fifth Street, New York 10017).

Index

U

V

W